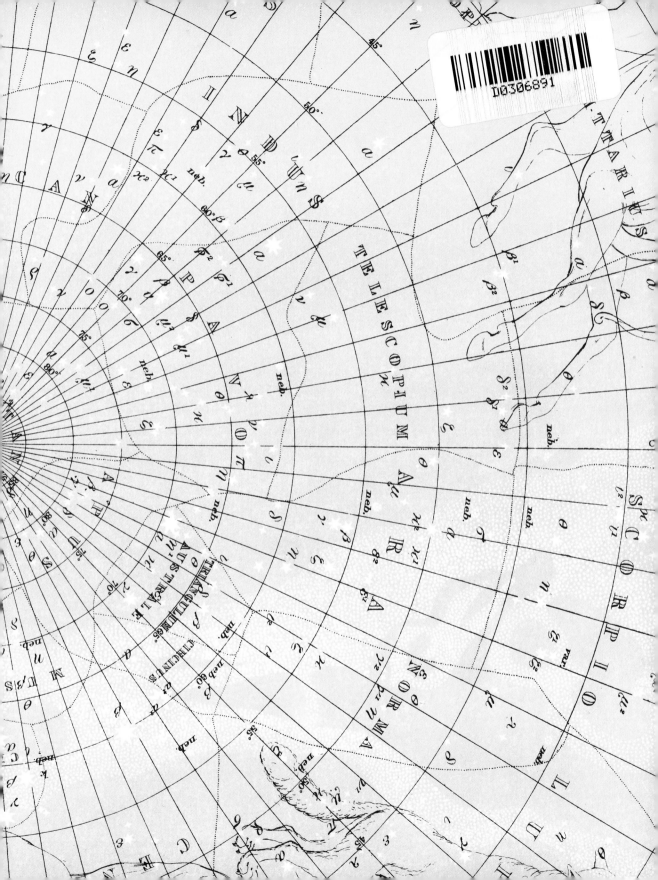

STEPHEN HAWKING'S
A BRIEF HISTORY OF TIME
A READER'S COMPANION

STEPHEN HAWKING'S

A BRIEF
HISTORY
OF TIME

A READER'S COMPANION

........———•———.......

Edited by Stephen Hawking

Prepared by Gene Stone

BANTAM PRESS

London • New York • Toronto • Sydney • Auckland

The documentary film *A Brief History of Time* is produced by
Anglia Television/Gordon Freedman Productions

Production design (and matte drawings):	Ted Bafaloukos
Director of Photography:	John Bailey, A.S.C.
Additional Photography:	Stefan Czapsky
Music composed by:	Philip Glass
Editor:	Brad Fuller
Executive Producer (Anglia Television):	Colin Ewing
Executive Producer (Gordon Freedman Productions):	Gordon Freedman
Produced by:	David Hickman
Directed by:	Errol Morris

TRANSWORLD PUBLISHERS LTD
61–63 Uxbridge Road, London W5 5SA

TRANSWORLD PUBLISHERS (AUSTRALIA) PTY LTD
15–23 Helles Avenue, Moorebank, NSW 2170

TRANSWORLD PUBLISHERS (NZ) LTD
3 William Pickering Drive,
Albany, Auckland

Published 1992 by Bantam Press
a division of Transworld Publishers Ltd
Copyright © 1992 by Anglia Television Inc./Gordon Freedman Productions

The endpaper star map is reproduced from *The Twentieth Century Atlas of Popular Astronomy* by
Thomas Heath (third edition, 1922), published by W. & A. K. Johnston, Ltd., Edinburgh,
and is provided through the courtesy of Isobel Hawking.

Book design by Michael Mendelsohn of M 'N O Production Services, Inc.

The right of Stephen Hawking to be identified as author of this work has been asserted in
accordance with sections 77 and 78 of the Copyright Designs and Patents Act 1988.

A catalogue record for this book is available from the British Library.

ISBN 0593 025105

Printed and bound in Great Britain by
Mackays of Chatham PLC, Chatham, Kent

*W*hich came first, the chicken or the egg?

Did the universe have a beginning,

and if so, what happened before then?

Where did the universe come from,

and where is it going?

STEPHEN HAWKING

FOREWORD

*M*y main aim in writing my book, *A Brief History of Time,* was to tell people about the progress that was being made in understanding the laws that govern the universe. I thought that others would share the excitement and the sense of wonder that I had felt if the basic ideas could be explained in a simple way. And that means without equations. Each equation, I was told, would halve the sales of the book. But that was okay. Equations are necessary if you are doing accountancy, but they are the boring part of mathematics. Most of the interesting ideas can be conveyed by words or pictures.

Of course, I hoped that the book would be successful, and that it would bring in a modest amount of money. At the time I began writing it, in 1982, I was thinking of my daughter's school fees. But I never expected the book to do so well. Since it was first published on April Fool's Day 1988, it has been translated into thirty languages and has sold about five and a half million copies worldwide. That is about one copy for every nine hundred and seventy men, women, and children in the world. Why have all these people bought it? There have been a number of attempts at explanation. It has been suggested that people buy the book but don't actually read it, far less understand it. It is said that they just want to be seen with the book, or that owning it gives them the comfortable feeling that they are in possession of knowledge, without their having to go to the effort of reading it.

I'm probably not an objective judge, but I don't believe this is the whole story. Wherever I go, all over the world, people come up to me and say how much they have enjoyed my book. These are ordinary people, not the trendy set or science freaks. Most of them seem to have read it, sometimes several times. They may not have understood everything they have read. If they had, they would be ready to start a Ph.D. in theoretical physics. But I hope they get the feeling that they are not shut out of the really big questions, and that if they work hard, they could understand more. I think some critics are rather patronizing to the general public. They feel that they, the critics, are very clever people, and if they can't understand my book completely, then ordinary mortals have no chance.

Although five and a half million is a great success for a book, it is still only reaching a small proportion of the population. The only way to reach a larger audience is through films and television. I was therefore receptive when Gordon Freedman approached me to make a film about six months after the book was first published. I had envisaged a documentary-style film that would be almost exclusively about science, with a large number of graphics. However, when they began to make the film, it seemed it was going to be almost entirely about my life, with very little about science. When I protested, I was told that the kind of film I had in mind would have appeal only to a fairly small number of people. To reach a mass audience, it was necessary to combine the science with material about my life. I was dubious. I thought it might just be an excuse to do a film biography, something I had earlier rejected. But working with the director, Errol Morris, has convinced me that he is a man of integrity, something rather rare in the film world. If anyone could make a film that people would want to watch, but which doesn't lose sight of the purpose of the book, it is he.

This book, *A Brief History of Time: A Reader's Companion,* is designed to provide background for those who read the book or watch the film. It contains a lot more material than could be included in the film, together with photographs from the film and explanations of the scientific ideas involved. It is The Book of The Film of The Book. I don't know if they are planning a Film of The Book of The Film of The Book.

STEPHEN HAWKING
Cambridge, January 1992

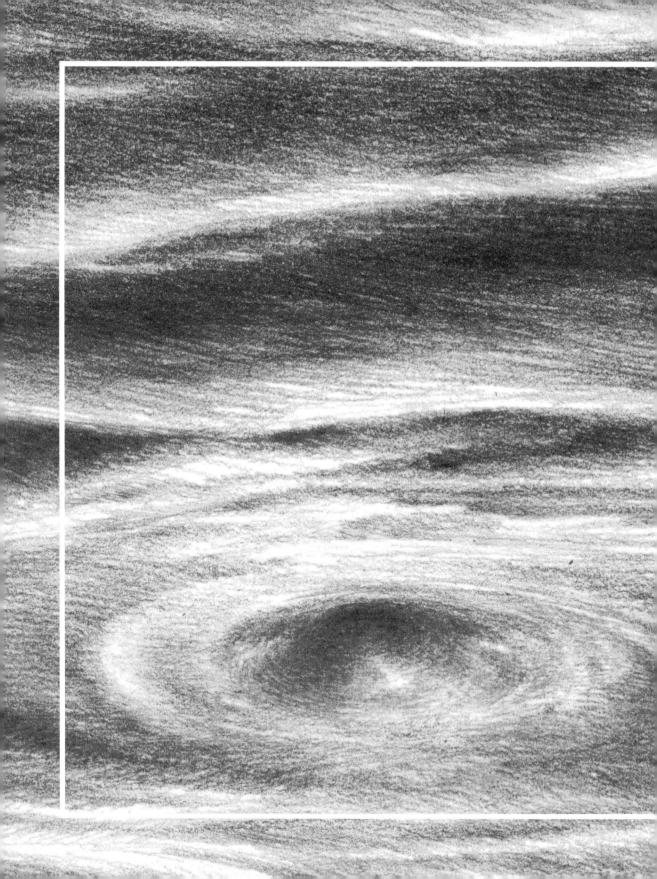

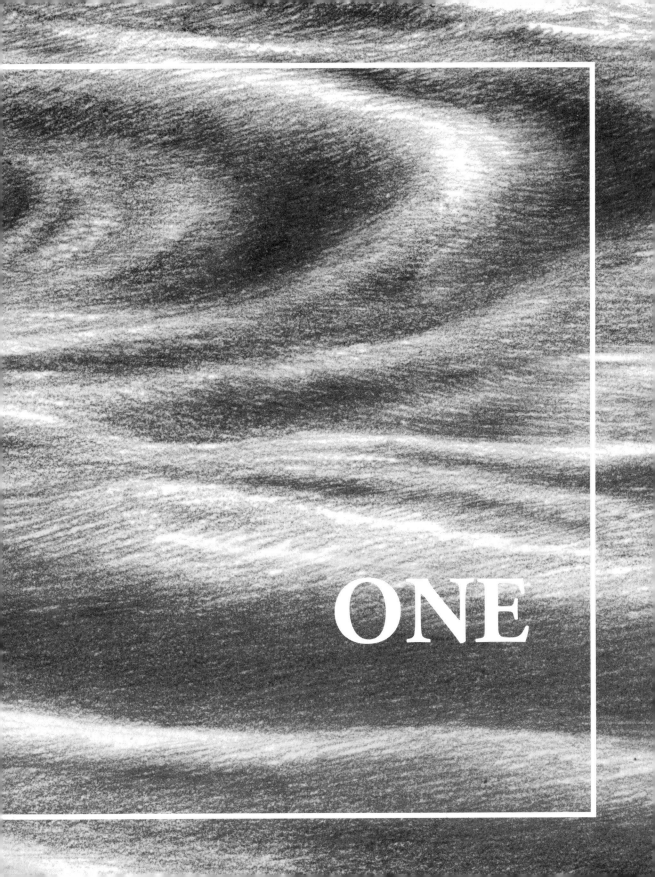

ONE

In January of 1942, when Frank and Isobel Hawking were expecting their first child, squadrons of Nazi aircraft were showering bombs over England's cities. Almost nightly raids on London caused the Hawkings to leave their home in Highgate for Oxford, to ensure a safe haven for their baby's birth.

They returned to London after their son, Stephen, was born, and lived there until 1950. Then they moved to the cathedral city of St. Albans, twenty miles north of London, where they raised Stephen, Mary (born in 1943), Philippa (1946), and Edward (1955).

Isobel Hawking is Stephen Hawking's mother. Now in her late seventies, she was one of seven children, and her husband, Frank, was one of five—at the last Hawking family reunion, eighty-three members showed up. Frank Hawking, who died in 1986, was a physician and a research biologist in tropical diseases for the National Institute for Medical Research. Isobel studied philosophy, politics, and economics at Oxford in the 1930s.

ISOBEL HAWKING

Luck. Luck. Well, we have been very lucky. I mean, my family and Stephen and everybody. You have your disasters, but the point is that we have survived. Everybody has disasters, and yet some people disappear and are never seen again.

Flying bombs are very alarming. They came buzzing over and then they would cut out. Then you started counting—they took, I've forgotten how long to fall. But when you heard the bang you knew it wasn't you, so you went back to your meal or whatever.

But one did fall quite close to our house. It blew the back windows out, so that the glass was sticking all dagger points out of the opposite wall.

So we decided Stephen had better be born in Oxford, and I went down a week before. First we went to a hotel, but they said, "You can't stay here—you might give birth any minute," so I had to move to the hospital. While I was there, I was doing this and that, and I had a book token, so I went to Blackwell's in Oxford. I bought an astronomical atlas.

One of my sisters-in-law said, "This is a very prophetic thing for you to have done."

STEPHEN HAWKING

I was born on January the 8th, 1942, exactly three hundred years after the death of Galileo. However, I estimate that about two hundred thousand other babies were also born on that day. I don't know whether any of them were later interested in astronomy.

I was born in Oxford, even though my parents were living in London. This was because Oxford was a good place to be born during the war: the Germans had an agreement that they would not bomb Oxford and Cambridge, in return for the British not bombing Heidelberg and Göttingen. It is a pity that this civilized sort of arrangement couldn't have been extended to more areas.

My father came from Yorkshire. His parents had gone bankrupt at the beginning of this century but had managed to send him to Oxford, where he studied medicine. He then went into research in tropical medicine. My mother was born in Glasgow, Scotland, and like my father's family, they were not well off. Nevertheless, they managed to send her to Oxford.

After Oxford she had various jobs, including Inspector of Taxes, which she did not like and gave up to become a secretary. That was how she met my father in the early years of the war.

I was a fairly normal small boy, slow to learn to read, and

very interested in how things worked. I was never more than halfway up the class at school. (It was a very bright class.) When I was twelve, one of my friends bet another friend a bag of sweets that I would never come to anything. I don't know if this bet was ever settled, and if so, which way it was decided.

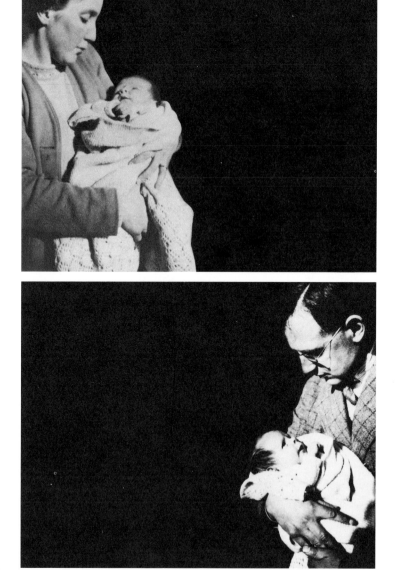

Isobel Hawking with Stephen, 1942

Frank Hawking with Stephen, 1942

JANET HUMPHREY

*M*y first memory is of Isobel pushing a rather antiquated carriage-built pram along North Road with Stephen and Mary in it, looking very large, because they had large heads and pink cheeks and they were very noticeable. They all looked different from ordinary people.

Janet Humphrey was trained in general medicine, then in psychiatry, and is currently a practicing Freudian analyst. Her husband, John, worked at the same institute as Stephen Hawking's father. The Humphreys met the Hawkings when Simon Humphrey and Stephen Hawking were enrolled at the same primary school in Highgate. The two boys became close friends. In 1959, when the rest of the Hawkings went to live in India, Stephen spent the year with the Humphreys while attending St. Albans School.

ISOBEL HAWKING

*S*tephen was certainly a very advanced child in some ways, but not in all. He didn't learn to read very early; his sisters learned much more quickly. But he was always extremely conversational; he was also very imaginative, and that side of him was brought out more than the mathematical side. And he loved music and acting in plays. One of the things he remembers most clearly is being taken to what must have been the first performance of Benjamin Britten's *Let's Make an Opera*. But Stephen never developed his musical side because I think he was rather lazy, and he had a lot of other things he liked doing.

Basically, they were just children, and we were mostly concerned with my husband's brilliance rather than Stephen's. Still, Stephen was a self-educator from the start, and if he didn't want to learn things it was probably because he didn't need to. He was generally like a bit

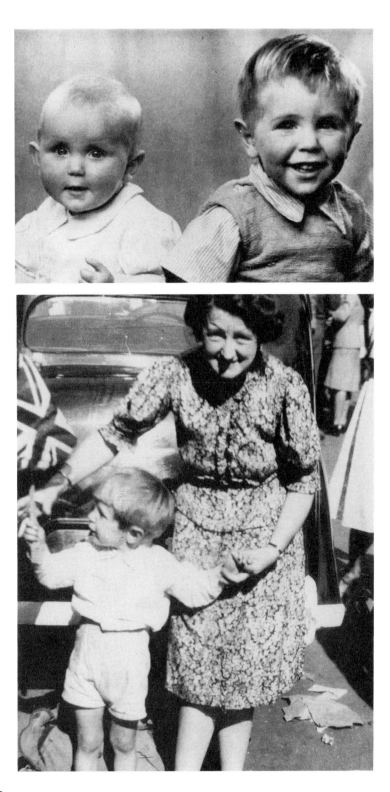

Stephen and his sister Mary, 1946. "They had large heads and pink cheeks and they were very noticeable."

Three-year-old Stephen on V-J Day, August 1945, with his Aunt Muriel, Frank Hawking's sister

of blotting paper, sort of sopping it all up. We used to take him and his sisters to the museums down at South Kensington. I would leave him in the Science Museum, Mary in the Natural History Museum, and Philippa, who was from an early age extremely aesthetic, I would take to

Stephen and Mary playing at the beach, 1946

the Victoria and Albert—I would stay with her because she was so young, but the others I left by themselves. Later I would go around and collect them. None of them would ever dream of going to the others' museums—they were just totally different.

Stephen was never much influenced by his father's sort of work; he was never interested in biology, he didn't want pets. From the start he was making things, thinking things, and talking a lot.

Mary Hawking, eighteen months younger than her brother Stephen, studied medicine at St. Barts in London and is now a general practitioner in Dunstable.

MARY HAWKING

*S*tephen used to reckon he knew, I think it was eleven ways of getting into the house. I could only find ten of them. I still don't know what the eleventh one was. On the north side of the house was a bicycle shed. It had a door at the front and a door at the back. Above that there was a window into the L-shaped room. And at the front you could get sort of around the corner onto the roof, and from that level you could get onto the main roof. I think one of the ways Stephen could get in was on the main roof. As I say, he was a much better climber than I was. We're not sure where the other ways were. They couldn't have been over the porch. The porch was pretty rotten even then, and there was an awful lot of glass about. Behind the porch was the greenhouse and that was practically falling down even then. Every time there was a wind you lost a few panes of glass.

EDWARD HAWKING

*T*he most striking thing about the house was the fence. Several times I tried to persuade my father to pull the fence down and let the hedge grow, but he insisted on patching this fence up. So, rather than spend any money, he would find any odd bits of wood that he had lying around and sort of nail them up against the fence.

When I took friends there I was a bit embarrassed. The front door was once very elegant. It had stained glass, but some of the glass had broken. Rather than replace the glass he used to get bits of body filler or putty and just slap it on. The wallpaper, although it was very grand, was also an embarrassment. It had been there for God knows how long.

It was a very large, dark house. And it was really rather spooky, rather like a nightmare. I had the sort of room that,

Edward Hawking is almost fourteen years younger than his brother, Stephen. He runs a small building firm in Luton, about thirty miles north of London.

The Hawking house in St. Albans

10

The "once very elegant" front door of the Hawking home

when you woke up on a winter morning, had thick frost on the inside. There was one radiator which didn't work properly, which was replaced by a storage heater in the hall. All the bedrooms had fireplaces, but of course it wasn't practical to have a fire in each room, so we just had the one fire downstairs.

The whole house was perhaps a little bit like the Munsters. But you know, it was our house and we loved it anyway.

ISOBEL HAWKING

We used to go to a pantomime at Christmastime. Once they had *Aladdin*, and at one stage Aladdin's palace was lifted magically and flew off into the air.

When we came out of the theater it took us a long time to get home because Stephen had to go and look for the palace. He knew then that what goes up must come down. Somewhere in Hampstead this palace was to be found. It took us a long time to persuade him this wasn't so.

The house Stephen imagined in "Drane": Kenwood House on Hampstead Heath.

He also had an imaginary house which he used to tell me was in a place called Drane. He used to have a tendency to try to leap onto buses to go there. We had to restrain this.

And when we went once to Kenwood House, up on Hampstead Heath, Stephen suddenly realized that this was his house in Drane. He told me in quiet tones that this in fact was it—apparently he had dreamed it or something!

Janet Humphrey

Stephen was very engaging, very lively and emotional, and his words couldn't keep up with his thoughts. So he sort of stuttered along sometimes when he talked. He was the same age as my son Simon, but he was smaller. I remember once they were coming back along North Road from school and some other boys started teasing, and it was Stephen who turned around and clenched his fists and threatened them, despite his size. But that's the sort of thing he did—he was equal to anything.

12

ISOBEL HAWKING

*T*he first year he was at St. Albans School he came in, I think, third from the bottom. I said, "Stephen, do you really have do as far down as all that?" And he said, "A lot of other people didn't do a lot better!" He was quite unconcerned.

He didn't do very well at school, but somehow he was always recognized as being very bright. In fact they gave him the Divinity prize one year, which was not surprising because his father used to read him Bible stories from a very early age. He knew them all very well. He was quite well versed in religious things, although I don't think he makes a very great deal of practice of it now.

MARY HAWKING

*F*ather was in tropical medicine. He used to do fieldwork, usually at the beginning of the year because that's the right time in Africa. So I always had the impression that fathers were like migratory birds. They were there for Christmas, and then they vanished until the weather got warm. The fact that everybody else's father seemed to be around at this time of year just convinced me that other people's fathers were a bit odd. He always came back with wonderful things—carved wooden animals and porcupine quills and pawpaws and things like that.

ISOBEL HAWKING

*M*y husband had a remarkably wide range of interests, and medicine was only one of them. Actually, the medical part wasn't really what interested him, anyway; he'd never have made a general practitioner. What interested him was research—it could have been research in almost anything but it happened to be in medicine, and the particular circumstances of his life guided him into tropical medicine. And he was lucky, too, because he got a fellowship in 1937 which took him to Africa for two years to study trypanosomiasis.

We were actually a single-parent family for a large part of the time because he continued to go to Africa for about three months every winter. So he and Stephen didn't see a lot of each other. But he did introduce Stephen to an interest in astronomy. I remember we all used to lie on the grass looking straight up through the telescope and seeing the wonder of the stars. Stephen always had a strong sense of wonder, and I could see that the stars would draw him —and further than the stars.

JOHN McCLENAHAN

I've known Stephen since we were both about ten or eleven. Some of my earliest recollections of the family and the house are of hearing Wagner rolling around the enormous living room. They were always very keen on Wagner. The other things I remember about his family are mostly about how unusual they were—in retrospect, perhaps even more unusual than I thought at the time.

It was perhaps only slightly less astonishing, also in retrospect, that Stephen turned out to be as very, very bright as he obviously

John McClenahan met Stephen Hawking at St. Albans School when he was ten years old; they lost touch when they studied at different universities, then became friends again at Cambridge. He received a doctorate in engineering and is now an administrator at King's Fund College in London.

is. There wasn't a lot of evidence of that in his early years at school. He was uncoordinated physically, and I suppose he always has been. And he was not bright on the sort of usual run of academic measures, or even unusually high-performing in an academic sense.

MARY HAWKING

We kept bees in the basement, and one day something had happened to the normal system of getting rid of spare queens. They just went on producing queens until we had about six or seven swarms. Mother had to keep collecting these swarms, not knowing where else to put them. Eventually she put some of them into an entrance outside the cellar, which seemed like a good place when you've run out of hives. But that was the night that our lodgers had locked themselves out and tried to get in through that entrance. Fortunately, the bees were very sleepy because it was dark.

BASIL KING

Stephen was the only person I knew at school who had to be issued with a copybook because his writing was so appalling. He was given a book with sentences in copperplate writing, with five or six lines under each model to copy out. How long he persisted with it or how long he was expected to operate with it, I don't know. But that is a sign his handwriting was considered spectacularly bad.

15

Basil King, a schoolmate and close friend of Stephen Hawking's at St. Albans, is now a doctor specializing in pediatric tropical medicine. He works for an international charity in Kenya.

I can remember visiting the Hawking home several times. It was the sort of place where, if you were invited to stay to supper, you might be allowed to have your conversation with Stephen, but the rest of the family would be sitting at the table reading a book—a behavior which was not really approved of in my circle, but which was tolerated from the Hawkings because they were recognized to be very eccentric, highly intelligent, very clever people—but still a bit odd.

I have a very clear memory of Frank Hawking, Stephen's father, sitting in his dressing gown over all his ordinary clothes in front of a closed combustion stove, in an attempt to keep warm. Frank Hawking had an alarming stutter. The belief amongst the rest of us was that the Hawkings were so clever that their speech couldn't keep up with their thoughts; that's why they stuttered, why they stumbled, why they ran over themselves in this rather clumsy form of speaking. I think you can see that in other members of the family as well. And something of that could be seen in Stephen, too, at the time.

John McClenahan

The house was crammed with books and bookshelves. Most of the shelves had books two deep, books horizontal across the tops of the ones that were in the bookshelves.

Stephen's father was, for me at least, a fairly distant figure. I think he was very shy. Children were difficult for him to deal with, whether they were his own or anyone else's. He seemed to live on a slightly different plane, thinking quite a lot about work, looking after a large and straggly house, and trying to bring up a moderately sized family on a not very large income, I think, for the time.

Stephen's mother was warmer, although still probably a little bit shy. My impression was that she largely ran the house, and ran the children, since Stephen's father was away on overseas trips a fair bit.

I always felt welcome there; I lost count of the number of dinners we had at each other's house. It was something that would just happen at the spur of the moment—we didn't need an invitation. We might both come home from school and get engrossed in a conversation and just decide that it wasn't worth going home.

This was also interesting for me because Stephen's mother was a much more adventurous cook than my mother, so I remember vividly having things like kedgeree for the first time—it's not that unusual nowadays, but back then it wasn't something I'd ever come across.

ISOBEL HAWKING

We had a caravan quite early on, on a field at Osmington Mills. It was beautiful, although when we bought it we thought we were never going to be able to use it at all because it was full of bugs. It had a double skin, you see, and all the bugs had gotten in between the two skins. But we had it fumigated, and it never had any more bugs.

So we put it on this field. We also had a huge army

17

tent to overspill, and we used to spend nearly all our holidays down there for a number of years. The children were very happy there—we could walk just about a hundred yards down to a rocky beach.

Such was our addiction to the wild that we took the children down to the caravan for the Queen's coronation in 1953. Apparently they've held it against us ever since,

Stephen, Mary, and Philippa in front of the caravan, August 1952

because we were depriving them of this great national experience when everybody else was having street parties and get-togethers, which we were never very good at. Mary says that she was receiving the coronation on a radio when my husband said, "Come along now, it's time to go to the beach."

So the children were rather forced into this wilderness, perhaps more than they wanted.

We used to get there in a taxi—we bought a succession of London taxis to go here and there; this was before cars were at all available. You bought them secondhand and you put a table in the middle, two children here, two children there, and they could play cards. They could do anything all the way there.

JOHN McCLENAHAN

*W*e made various expeditions in the Hawking family car of the time, which was an old London taxi—a very old London taxi, just postwar. Once we went sailing along the open countryside towards Maiden Castle, with Stephen's mother driving in the front, and three or four children, including me, in the back, all looking over this open-top taxi. The front didn't have a cover and the back had a cover folded down, so we were entirely open to the air, barreling along at about forty miles an hour, which was very nearly as fast as the thing would go.

It was the kind of family that, it seemed to me, did those sorts of odd things. We didn't have a car. Most people in Britain at that time didn't, unless they were fairly wealthy. And to have a car that was a beaten-up battered old London taxi was even more unusual.

MARY HAWKING

I have a very clear picture in my mind of Stephen absolutely buried in some book, I don't know what it was, with a tin of biscuits beside him. And you couldn't get his attention. He was totally absorbed in this book and the biscuits just sort of vanished as though by . . . well, I think he was quite surprised when he came out of the book and found the biscuits had all gone!

ISOBEL HAWKING

*E*ven as a small boy, if he was interested in something, he would concentrate on it one hundred percent. I remember him sitting in, I think it was a tractor or some agricultural thing of his farming relations, studying what was what, and the other children were actually climbing over him, sort of walking on his head, you know, but he took no notice at all.

MICHAEL CHURCH

I first met him when we were in the third form at school. He was one of the bright boys in the class, one of the six or eight clever ones. He wasn't top, he was just one of the top set. He was disheveled, with an ink-stained collar—fun to be with, but also physically puny. He tended to get bashed around in the showers, tended to get picked last for any team. But it didn't

Michael Church met Stephen Hawking at St. Albans School in 1957. The two lost contact while Church studied arts at Oxford, and then renewed their friendship later in life. Church recently left a job as a newspaper correspondent on The Independent *in London and is now a free-lance journalist.*

seem to bother him too much. He enjoyed himself.

He spoke very fast—he was almost incoherent. And he had a special kind of language, a way of talking that collapsed words, sometimes quite creatively. I remember once he talked not about "silhouettes" but about "slit-outs," which is actually quite an interesting collapsing of the word.

ISOBEL HAWKING

When he was thirteen he did have an illness which may have had something to do with his later one. I don't suppose we shall ever know. As far as it was diagnosed at all, it was diagnosed as glandular fever; it was a low fever that kept recurring, and it went on for a long time. It was subacute most of the time, but sometimes he was actually in bed with it, and he was off school quite a long time. Then he just sort of got better, but whether he was quite the same I don't know.

MARY HAWKING

Father was very good at theological debate, so everybody used to argue theology. A good safe subject. You don't need any facts or distracting things like that. If you go in for arguing, you can quite happily debate about anything— including theology, and the existence or otherwise of God. And then someone gets bored or "Journey into Space" comes on or something like that, and the argument breaks up.

21

JOHN McCLENAHAN

*S*tephen's father had a conservatory greenhouse, and it was in the greenhouse that we used to make fireworks. I'm not sure where the recipes came from, and I suspect in hindsight that some of them were very dangerous.

Once we had a quite well liked but very stern English teacher who was teaching us Shakespearean plays, and on April first a group of us decided we wanted to make him a little less stuffy than he appeared. We made potassium iodide, which dried on filter paper into a percussive explosive. We then put a piece under each leg of his chair so that as soon as he came in and sat down they would go off. And they did. He was more than a little startled by it. We also had written up on the board where he could see it when he turned around the line: "Dost thou think, because thou art virtuous, there shall be no more cakes and ale?," which was a quotation from *Twelfth Night*.

He took it extraordinarily well, bless his heart.

ISOBEL HAWKING

*F*ireworks were a bit scarce and also expensive, so they used to make their own. My husband, of course, was perfectly in control of the situation, so it was quite safe, but still, I didn't like it.

They made them in a shed and they let them off on Guy Fawkes' Day. The children learned quite a lot about chemistry this way, such as the different colors you put in to produce the different effects. And the fireworks were quite effective; both Stephen and his father enjoyed them very much indeed.

Stephen and his father also went surveying together. I think everybody ought to go surveying, because it's practical and theoretical and also very beautiful. They used to go out to Ivinghoe; we used to go a lot to Ivinghoe Beacon, on the Chilterns. And they would go along one of the paths and they would survey things here, there, and everywhere a long way off, and make notes together.

MICHAEL CHURCH

I had not taken him seriously at all, because he was just a bright little boffin, and a great friend, of course, but not somebody with any great perception or any great understanding of the meaning of life and what it's all about. Then one afternoon we were messing about in his boffin's den, which was chaotic and generally a joke, a sort of joke scientist's den, and we began talking about life and philosophy and so on, which I thought I was very hot on at the time, and so I held forth.

Suddenly I was aware that he was egging me on, leading me to make a fool of myself. It was an unnerving moment. I felt looked down upon from a great height. I felt that he was watching, amused and distant.

It was at this point that I realized for the first time that he was in some way different and not just bright, not just clever, not just original, but exceptional. And there was some overarching arrogance, if you like, some overarching sense of what the world was about.

23

BASIL KING

We were discussing the possibility of the spontaneous generation of life. I think that Stephen made a remark which indicated not only that he'd thought of this, but he had even come across some calculations as to how long it might take. At that time I made a comment to one of my friends, John McClenahan: "I think that Stephen will turn out to be unusually capable."

John disagreed. So in our childish way we bet a bag of sweets on the issue. And, incidentally, I reckon that my bet has come correct and I'm entitled to payment, which has not yet been made.

JOHN MCCLENAHAN

Well, three of us had a bet that none of us would amount to much, or that some of us would. I forget which way round the bet was, but Stephen still maintains that I owe them a bag of sweets because I've never delivered on him having become famous.

It's very difficult to tell, even with hindsight. He was very unusual, but not obviously theoretically brilliant. Though there is one story—I'm not sure why we were discussing it, but the question was, if you have a cup of tea which is too hot to drink, will it cool more quickly if you put milk in first, or leave the milk out and put the milk in later? I couldn't myself think of how to address this question at all. But for Stephen it came just like that—a flash of illumination. I can reproduce the argument: any hot body loses heat at a rate proportional to the fourth power of its absolute temperature. Stephen said that therefore the longer you leave it undiluted by the milk the faster it will cool, and therefore you put the milk in at the end, not the beginning.

MARY HAWKING

I gave up playing games with Stephen when he was about twelve because he started taking games terribly seriously. We had Monopoly. First of all, to add to the complications, the Monopoly board sprang railways going across it. Then Monopoly just wasn't adaptable enough, and he ended up with a fearful game called Dynasty, which, as far as I can make out—as I say, I never played it—went on forever because there was no way of ending it.

ISOBEL HAWKING

The game was almost a substitute for living as far as I could make out. It took hours and hours and hours. I thought it was a perfectly terrible game; I couldn't imagine anyone getting taken up with that. But Stephen always had a very complicated mind, and I felt as much as anything it was the complication of it that appealed to him.

JOHN MCCLENAHAN

Stephen was very adept at inventing these complex board games. Monopoly was child's play by comparison. The games were played on a very large piece of cardboard, probably three feet by two or more, divided into half-inch squares. They were mostly war games, with complex rules about how far you could move on the throw of a dice. A typical game would last at least four or five hours, and several of them would run for many sessions over a week.

MICHAEL CHURCH

*H*e loved devising rules. One of his biggest achievements was a feudal game where one throw of the dice going around the table took an entire evening to work out the consequences. It was labyrinthine. He loved the fact that he had created the world and then created the laws that governed it. And that he was causing us to obey those laws: he enjoyed that, too.

ISOBEL HAWKING

I think they were in the fifth form when they built the computer. Yes, it must have been the fifth, because in the

Stephen with his bicycle, 1957

26

sixth form they'd probably all have been too busy. There were, I think, six of them. This was in 1957 or '58, the early days for computers. They made it out of a very great collection of this and that, such as the insides of clocks and various things. And it did answer questions. We all went to see it at the school, where it was quite a sensation. As long as you asked the right questions, you quite likely got the right answers.

Now it wasn't just Stephen's, because he was never good with his hands. He would be, perhaps not the only, but one of the brains behind it—I think a lot of the practical work was done by John McClenahan, who was very good with his hands. Anyhow, they all worked together.

JOHN MCCLENAHAN

When we were making these sorts of computers and things, I remember he could really do the more complex maneuvers, but I could manage things that he would have a go at and get back to me and say, "I can't do this." I have a general impression of gangling. But I have other friends who were very much the same and who haven't gone on to suffer his disease. A university friend of mine was just as gangling back then, and he isn't ill now. In Stephen's case, then, I think it was more a case of nervous and physical mannerisms.

ISOBEL HAWKING

In his last year at St. Albans School we had to go to India, because my husband had a job with what was then called the Colombo Plan, through which scientists and others were sent out to institutes in India and elsewhere, to work with the people there, to exchange knowledge and so on. So when Frank was appointed to this, we all went except Stephen. He was doing A Levels that year, and we didn't think he should come.

JANET HUMPHREY

*W*hen the Hawking family went to India, it was decided that Stephen would come live with us for a year. We had a large house and a large family, and it was a time when he couldn't afford to leave—one couldn't just go away from school; it was too important a year to leave out. And of course he would be quite all right with us.

Stephen was rather clumsy in his movement. I remember him running with a trolley full of crockery into the kitchen, having cleared the table, and bumping into something and the whole lot came off. I suppose everybody laughed, but after a pause Stephen laughed the loudest.

But at the same time he was rather organized: for instance, he once decided it would be nice to have Scottish dancing in the evening. Now mind you, this was quite an ordinary house, but we had rather a lot of room and a large hall. We bought some records and a book about what to do, and Stephen took charge. He insisted that you put on a jacket and a tie, because he was one of the oldest of the children. And he was the master of the proceedings.

I don't know how often we did it, but it certainly was a great pleasure. Stephen took it very seriously—but then, he liked dancing, you see.

ISOBEL HAWKING

 I still have the vigorous correspondence that we all had with Stephen during that time, because although Stephen was a natural thrower-away of letters, the Humphreys made him keep our letters. Unfortunately, I can't find Stephen's, which I expect were of a remarkable dullness because he wasn't ever much of a letter writer. I think he answered the letters simply because Dr. Humphrey sat him down and said, "Now you write letters to your family."

But he had a very good year with the Humphreys, and we had a marvelous time in India. Stephen joined us at the very end, when he had already taken his Oxford exam and got his scholarship.

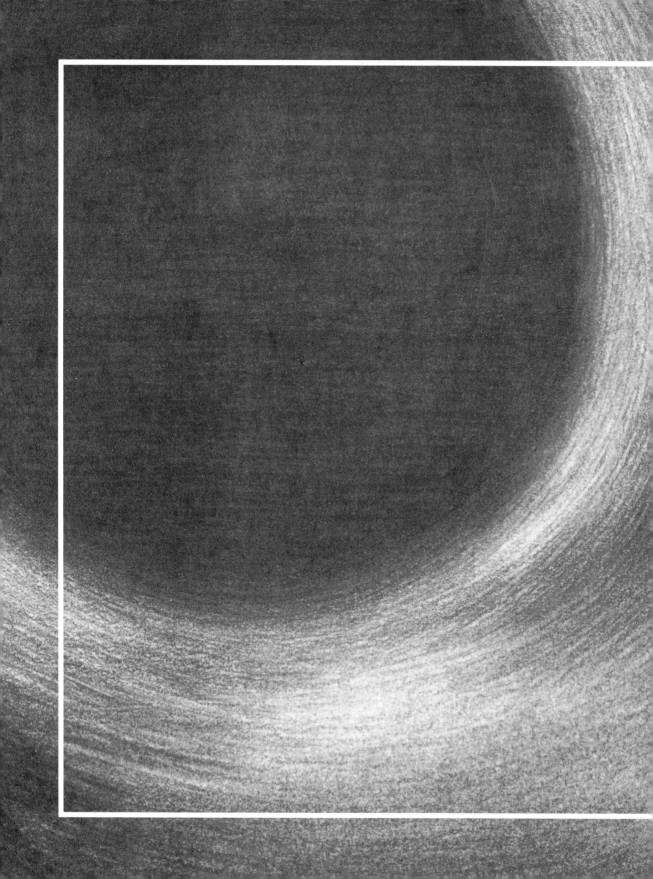

TWO

STEPHEN HAWKING

My father would have liked me to do medicine. However, I felt that biology was too descriptive and not sufficiently fundamental. I wanted to do mathematics and physics. My father felt, however, that there would not be any jobs in mathematics apart from teaching. He therefore made me do chemistry, physics, and only a small amount of mathematics.

Another reason against mathematics was that he wanted me to go to his old college, University College, Oxford, and they did not do mathematics at that time. I duly went to University College in 1959, to do physics, which was the subject that interested me most, since it governs how the universe behaves. To me, mathematics is just a tool with which to do physics.

Most of the other students in my year had done military service and were a lot older. I felt rather lonely during my first year, and part of the second. It was only in my third year that I really felt happy at Oxford. The prevailing attitude at Oxford at that time was very antiwork. You were supposed either to be brilliant without effort, or to accept your limitations and get a fourth-class degree. To work hard to get a better class of degree was regarded as the work of a gray man, the worst epithet in the Oxford vocabulary.

At that time, the physics course at Oxford was arranged in a way that made it particularly easy to avoid work. I did one exam before I went up, and then had three years at Oxford, with just the finals exam at the end. I once calculated that I did about a thousand hours' work in the three years I was at Oxford, an average of an hour a day. I'm not proud of this lack of work, I'm just describing my attitude at the time, which I shared with most of my fellow

33

students: an attitude of complete boredom and feeling that nothing was worth making an effort for.

DEREK POWNEY

Derek Powney was one of the four physics students at University College in Stephen Hawking's year; after leaving Oxford, he did research at Bristol University and is now headmaster of the Abbs Cross School in Essex, just outside London.

*T*here were four physicists in my year at University: Stephen, Gordon Berry, Richard Bryan, and myself. The first cameo of Stephen I can remember is when Gordon and I went up to his room after dinner to try and find him. There he was, sitting with a crate of beer, slowly drinking his way through it. At that time he was only seventeen; he couldn't legally go into a pub, of course. Because he'd gone up to Oxford ridiculously early—he'd gone for a dry run on the scholarship exam the year before he should really have taken it, and to the school's amazement they'd given him a place, and so he'd gone up the following October.

I don't think at that time any of us knew just how intelligent Stephen was. We didn't realize it until our second year. We had been working in two pairs in tutorials, and for once all four of us were doing the same work—the two pairs were in exactly the same position. We were asked to read a chapter, chapter ten, in a book called *Electricity and Magnetism* by Bleaney and Bleaney, an unlikely combination, a husband-and-wife team. At the end of that chapter there were thirteen questions, and all of them were final honors questions. Our tutor, Bobby Berman, said, "Do as many as you can."

So we had a go and I discovered very rapidly that I

Stephen Hawking

Gordon Berry

Richard Bryan

Derek Powney

couldn't do any of them. Richard was my partner, and we worked together for the week and managed to do one and a half questions, of which we felt very proud. Gordon refused all assistance and managed to do one by himself. Stephen, as always, hadn't started. Stephen didn't do very much work when he was up.

We said to him, "It's no good, Hawking, you have to get up for breakfast in the morning." This would indeed be an event of its own, because he didn't get up for breakfast. He looked at us thoughtfully, and the next morning he did get up for breakfast. Being good little boys, we trotted off to our three lectures of the morning, whilst Stephen didn't. He went up to his rooms at nine o'clock, or five to nine, say, because it was only five minutes up to the labs from University.

We came back about twelve and down came Stephen. We were in the college gateway, in the lodge.

"Ah, Hawking!" I said. "How many have you managed to do, then?"

"Well," he said, "I've only had time to do the first ten." We fell about with laughter, which just froze on our lips because he was looking at us very quizzically. We suddenly realized that that's exactly what he had done—the first ten. I think at that point we realized that it was not just that we weren't in the same street, we weren't on the same planet.

Patrick Sandars was a Research Fellow in Physics at University College and one of Stephen Hawking's tutors. Today he is a Professor of Experimental Physics at the Clarendon Laboratory at Oxford.

PATRICK SANDARS

Stephen never showed any great interest in what he was set. Sometimes that lack of interest was stronger than at other times. Once we had to move to the topic of statistical physics, which was part of the theoretical option in the finals syllabus. I showed him the book that we were going to work through during the term. He took one look at it and seemed to take an instant dislike to it. Nonetheless, I persevered and said that we should take the first chapter in the first week and he should do a couple of the problems which I set him.

At the end of the week, at the next tutorial, rather than come back with the problems, he came back with the book, with every error marked in it but none of the problems done. He put it down and we had a brief discussion about the subject, by which time it was clear that he knew more about it than I did.

ROBERT BERMAN

Robert Berman studied at Cambridge before taking a position in the physics department at Oxford, where he was Stephen Hawking's supervisor and physics tutor at University College.

*W*hen I first met Stephen he was probably still not seventeen. His father, who had been an old member of the college, brought him to see me, and we talked generally about coming to college and reading physics and so on. In fact, as far as I remember, his father did most of the talking; I didn't really get much impression of Stephen.

But when he took the entrance exam, he did very well, especially in the physics. At the general interview, which in those days was held in the presence of the master and the senior tutor and various other college authorities, everyone immediately agreed he was absolutely an A character as a future undergraduate, so there was no question of not giving him a scholarship and taking him to read physics.

He was obviously the brightest student I've ever had. He didn't do as well in the final examination as probably thirty others I've had since then, but of course the ones who do better not only are clever people but have worked very hard. Stephen, however, was not just clever, he was more than clever; but it couldn't be said that by normal standards he worked very hard, because it wasn't really necessary. He produced his work every week for his tutorials, and I think that was really my role, just to keep him ticking over doing some work in physics. I'm not conceited enough to think that I ever taught him anything.

Gordon Berry was Stephen Hawking's physics tutorial partner at University College and one of his closest friends. He was also, like Hawking, a coxswain. After graduating, Berry moved to the United States to study for his advanced degrees. Until recently he held a joint appointment in the Department of Nuclear Physics at the University of Chicago and the Argonne National Laboratory; today he works only at Argonne, where he heads one of four research teams.

GORDON BERRY

*S*teve was probably the person I knew most of the time I was at University College. We had tutorials together, we were on the river six afternoons a week, we also played bridge together in the evenings.

We did have a really excellent physics tutor, Bobby Berman, whose area was thermodynamics. Bobby, I think, despaired that we'd ever learn anything because we just didn't apply ourselves, and Steve was right down there in not applying himself. There was absolutely no sense that we would learn a lot of physics as undergraduates. The important thing was to have a good time and socialize. I mean, that's why I came to Oxford. We were obnoxious to our own families. I think our tutors definitely despaired of us ever doing anything constructive. Still, I think that what we learned in thermodynamics affected my work later on in life, and it had a very significant role in Steve's study of black holes.

NORMAN DIX

The Boat Club was a major college activity at University College, and while Stephen Hawking was an undergraduate, Norman Dix was College Boatsman. He worked at Oxford for more than forty years and is now retired.

$\mathcal{W}$e used to have what we would call a gathering net, and that's how we collected him. We used to organize a beer party and various things like that to collar as many freshmen as we could get, you see, to get them to join the Boat Club.

But the question with Stephen was, should we make him a cox of the first eight or the second eight. You see, some coxes can be adventurous and some coxes can be very steady, and Stephen was rather an adventurous type; you never knew quite what he was going to do when he went out with the crew. I think he used to bring his work with him onto the boat sometimes; he had his thinking gear going on different levels.

But he had a good, fairly clear voice—sometimes a voice that was not quite a sergeant major's, but a fairly commanding voice.

We had a very lively Boat Club at that time. They enjoyed the river and that's what it's all about. They didn't go and make vast inquests on things, rather like they do now. The terrible thing now is that in modern rowing, no one likes losing. They all get together and they sort of need a psychiatrist to examine their reasons for losing. I feel the reason you've lost is because someone is a little better than you.

GORDON BERRY

$\mathcal{M}$y memory's a little vague on how we got involved in the Boat Club, but I think they had a party in the cellars

Stephen Hawking coxing his boat, Eights Week, Oxford, 1961

and we went down there and had drinks. They would try to persuade the biggest guys in college to come down to the river and try out, even though they'd not rowed before. And then they would look for some little guys. I wasn't too little but I was very thin, and I think Steve was just small enough to be a cox; he wasn't big enough to row. We were both urged to come down and try our hand at coxing, and so we did.

We stuck on the river all three years. It was certainly our life. We were supposed to take the practicals, the physics experimental labs, three whole days, from nine in the morning till three in the afternoon. But of course Steve and I had to be on the river every afternoon, six days a week. Something had to give, and it was definitely the experimental labs. Steve and I became real experts at collecting data

40

The University College Boat Club, 1962. An ebullient Stephen Hawking is at the right.

very, very rapidly—collecting the minimum amount of data and the maximum amount of data analysis, so that it looked as though we had done the real experiment. This required quite a little bit of mirrors—we had to convince the people who were grading the experiments that we'd done everything, although they knew that we hadn't been around the lab. We had to be very careful writing up our experiments. We were never cheating, but there was a lot of interpretation going on.

DEREK POWNEY

*S*tephen was so intelligent compared with the rest of us that we must have been quite difficult to live with. I think

41

that part of Stephen's strategy for living with people who are in fact quite unintelligent by his standards is to find defense mechanisms with which to cope—the same kind of thing which you get on a different plane with people who are less intelligent, and so on. But even in Oxford we must all have been remarkably stupid by his standards, and it's quite difficult to live all the time with people who are a lot stupider than you are. I think that therefore you tend

At a Boyle Society dinner. The Boyle Society was University College's physics society.

to become a very private person, and to build almost a caricature of yourself as a defense.

ISOBEL HAWKING

*W*hen we saw him, he often seemed all right, because he wasn't doing anything out of the way. He wasn't walking with a stick, he was walking quite all right. Shortly after we came back from India, and before he went up to Oxford, I drove him and his younger brother to Woburn Park, where he climbed a tree. He was testing himself out, I think, but I didn't realize why. He did manage to climb a tree and to go along a branch of it, and to get himself down.

GORDON BERRY

*U*niv has these square staircases, which are round but they're square, and Steve actually fell on the stairs coming downstairs and kind of bounced all the way to the bottom, really quite a severe bump. I don't know if he lost consciousness—he may have done so momentarily—but he did lose his memory; he couldn't even remember who he was.

So we took him to my room, where he sat down on the couch, and his first question, of course, was "Who am I?"

We told him, "You're Steve Hawking."

Right away, he asked again, "Who am I?"

"Steve Hawking," we said.

And then, after a couple of minutes, he remembered that he was Steve Hawking. He didn't remember that he'd asked the question "Who am I?" but he remembered his family life from a year ago. And then he asked, "Where am I?" and we told him, "You're at Univ. You just fell down the stairs," and so on. He would hear the answer, but he would ask a different question, and he would start saying, "Well, I can remember that I came up to Univ, you know, in 1959," and he'd remember that he took classes and courses, but he couldn't remember what happened a year ago, and then what happened a month ago. So we would try to tell him what had happened at University during this time, and he'd say, "Oh, yes, I remember that." And then he would remember events that had happened a month ago, and then events that had happened a week ago.

Then it was obvious to him, to all of us, that his memory was gradually coming back, and that we just had to be patient. So we would ask questions to see if he could remember up to a week ago.

We'd say, "Well, do you remember going down to the bar and having a drink on Sunday night?" or "Do you remember coxing on the river on Monday?" We'd ask him until he remembered, and then we'd go to the day after that. It took a little longer as we got closer to the end result. I think it took about two hours before he could remember actually falling down the stairs.

The question was, well, maybe he had lost some of his mind because of this. Steve decided, "I'll just try this out and make sure that nothing severe has happened. I'll take the Mensa test." So he went and applied to Mensa: this would prove whether he was still bright or not. Of course he passed the Mensa test. He got 200 or 250 or something. So there was no problem whatever there.

Derek Powney

*W*hen we got around to the final honors degree—well, it's a bit daunting, because the whole of your three years' work goes on exams taken morning and afternoon, morning and afternoon, through the four days. So the four of us then decided to go out for dinner each night at a different restaurant. We would not talk shop, we would just have dinner together.

That last night, I remember, three of us were in a state of deep depression. Stephen thought he hadn't got his first. Richard thought he hadn't even managed to scrape a third; I reckoned I hadn't got my second. Gordon was very cheerful because he thought he'd got a first.

So three of us were in a depression and one of us was extremely happy. In actual fact, we all got it wrong. Stephen got his first; I got my second; Richard got his third; and Gordon didn't get his first. He got a second instead. So all four of us had got it absolutely wrong.

Early the next morning I just packed my cases and left Oxford very hurriedly by train, the 9:10 train, because it had been a golden age at Oxford, and I wanted not to spoil it.

STEPHEN HAWKING

*B*ecause of my lack of work, I had planned to get through the final exam by doing problems in theoretical physics and avoiding any questions that required factual knowledge. I didn't do very well. I was on the borderline between a first- and second-class degree.

I had to be interviewed to determine which I should get. They asked me about my future plans. I replied I wanted to do research. If they gave me a first, I would go to Cambridge. If they gave me a second, I would stay in Oxford. They gave me a first.

46

ISOBEL HAWKING

𝓘n his third year I think he himself began to notice that his hands were less useful than they had been, and that things were getting a bit difficult. But he didn't tell us. I was very worried about him, but we put it down to examination nerves and a general sort of youthful difficulty, so we didn't tumble to it.

I think it was in the summer term, after he had done his finals, just before he came down, that he had a fall downstairs; he fell on his head. He was always falling on his head. I told him he must go and see a doctor to have his reflexes tested to see whether he'd done himself any damage. So he went to see a doctor.

They didn't find anything neurologically wrong with him; he was not yet diagnosed with motor neurone disease.

Then he wanted to go to Iran, to Persia, with a friend. In those days you had to be twenty-one before you were of age, and we were a bit worried about it, actually. But although he wasn't twenty-one he was really quite grown-up, and this friend he was going with was experienced— he had been to Persia before. The friend's parents wrote to us and said their son knew all about it and how it would be all right.

So we agreed, and he went off to, well, we called it Persia in those days. They had a very interesting time and Stephen wrote us a couple of letters. The last letter we got said that he was just about to leave Tehran and go to Tabriz. I'm not sure how long that letter took to arrive, but he gave us the date he was supposed to be home—he was coming back on the student train from Istanbul.

And then there was nothing more. Then there was a very serious earthquake between Tehran and Tabriz, and we just didn't hear any more of him for three weeks. We got in touch with the Foreign Office, and it was a dreadful time—no news at all—except that the Foreign Office said there was no mention of any British person having been involved in this earthquake. But we knew it had taken place just where Stephen was going in a bus.

Then, three weeks later, he turned up on the Lebanon frontier. It seemed his bus had been jumping about so much, Stephen didn't notice the earthquake, though he presumably went over the earthquake area. But by the time he got to Tabriz he was so ill that he had to get off the bus. His friend stayed with him, and they saw a doctor in Tabriz. But they still didn't hear about the earthquake, and nobody mentioned it because they were strangers and nobody would have thought it was of any great interest to them. So they just didn't know.

When he finally came home he looked very much the worse for wear. But that was not the cause of the illness. He'd been ill well before that. He just didn't know—or at least we didn't know. But it was a bad setback, and no doubt his illness got quite a lot worse.

MARY HAWKING

Once we were talking about how not being able to walk along a straight line was perhaps a sign of having had a few too many beers the night before. Stephen said that he'd tried it out and he never could walk along a straight line. He didn't think there was anything in that one.

⸻

In the autumn of 1962 Stephen Hawking went up to Cambridge, but his physical problems worsened over the next few months.

48

STEPHEN HAWKING

*S*hortly after my twenty-first birthday I went into the hospital for tests. They took a muscle sample from my arm, stuck electrodes into me, and injected some radio-opaque fluid into my spine, and watched it going up and down with X rays as they tilted the bed. I was diagnosed as having ALS, amyotrophic lateral sclerosis, or motor neurone disease, as it is known in England.

The realization that I had an incurable disease that was likely to kill me in a few years was a bit of a shock. How could something like that happen to me?

⸻

Amyotrophic lateral sclerosis is known in the United States as Lou Gehrig's disease, after the New York Yankee first baseman who died of it in 1941. The disease is responsible for the disintegration of the nerve cells in the spinal cord and brain that regulate voluntary muscular activity. The mind is not affected. Death usually results from the failure of the respiratory muscles and comes in the form of pneumonia or suffocation.

⸻

ISOBEL HAWKING

*I*t was a very cold year and the ice on Verulamium Pond was frozen. We all went skating. Stephen managed to skate fairly well, but then he and I were close together. He wasn't skating in a very advanced way, nor was I when it comes to that.

Then he fell, and he couldn't get up. It was quite obvious

something was seriously wrong. So I took him to a cafe to warm up and he told me all about it.

I insisted on going to see his doctor, because it seemed to me that however long you're going to live, there's probably something someone can do about it—at least, to make things easier for people.

Now I won't mention the doctor's name, but I got to see him at the London Clinic. He was rather surprised that I should bother to come round to see him. After all, I was only Stephen's mother! He was quite nice, though, and agreed to see me in a rather grand way. He said, "Yes, it's all very sad, a brilliant young man cut off in the prime of his youth, as it were."

But of course I said, "What can we do, can we give him physical therapy or anything that will help in any way?"

"Well," the doctor said, "actually, no. There's nothing I can do really. More or less, that's it."

Of course, when my husband knew about it, he had to get hold of doctors in a different line from his to get a diagnosis. They decided that it was one of two things—it could have been a tumor on the brain, which was operable, and if it wasn't that, it was this.

When it was diagnosed, they said, "He probably won't live more than about two and a half years."

DEREK POWNEY

*W*ell, Stephen was always very clumsy. But I don't think anyone thought anything of that. There was that one time, towards the end of his third year, when he fell down the stairs coming from the junior common room. But nobody thought anything particular about that.

Once, later, on a visit to Oxford, I was looking for

someone to have lunch with, but nobody was around. And then, lo and behold, Stephen walked through the door. He generously went off to buy the drinks, brought them over, and put them on the table. As he put his pint of beer down he spilt it.

"Heavens," I said, "drinking at this time of day!"

Then he told me that he'd been in hospital for two weeks where they'd done a series of tests and decided what was wrong with him. He told me very straight and flat that he was gradually going to lose the use of his body. And that they'd told him eventually he would essentially have the body of a cabbage but his mind would still be in perfect working order, and he would be unable to communicate with the rest of the world. He said that at the end only his heart and his lungs and his mind would still be working. After that, either his heart or lungs would give up. Then he would die.

He'd been told that the disease was incurable, that it was totally unpredictable, that it could stop for long and for short periods of time, that it never got better but it could just arrest, and that there was no idea whether he would die in six months or twenty years. But because he had developed the disease at a much younger age than most people, they suspected that it would be sooner rather than later.

I was totally devastated, which must have been utterly unhelpful as far as Stephen was concerned. I knew perfectly well that he had no faith, and to me that made it all the more difficult, because you must ask yourself, "Why me? Why this? Why now?"

But he just flatly accepted that this was what was going to happen to him. And as far as I can gather, at that point he started to do some work. Certainly, within about eighteen months or so he'd had a paper published by the Royal Society, making a slight correction to Professor

Hoyle's latest theory of gravitation, for which Hoyle subsequently thanked him in acknowledgment. That's when his career took off, still as a research student, before he'd taken his doctorate.

ISOBEL HAWKING

*E*dward was only a little boy and I don't think that he particularly noticed, actually, but because everybody else had their attention elsewhere Edward wasn't getting the attention he really needed at that age. So he suffered, although he's not a suffering sufferer. And Mary, because she was awfully close to it, being at the hospital when Stephen was ill, she suffered, too.

And my husband suffered very much. But we got over it, didn't we? Except for my husband, who probably would be dead anyhow. He didn't die of that. But everyone else is still alive.

I don't know if Stephen ever realized how much his father tried to do when he was first ill. Because of course he wasn't telling Stephen about all his efforts, and how he got in touch with people like Carleton Gajdusek, who was a Nobel Prize winner in a related illness called kuru which occurs in Borneo or somewhere. But kuru is spread by cannibalism —and there's not a lot of that going on in this country.

STEPHEN HAWKING

*M*y dreams at that time were rather disturbed. Before my condition had been diagnosed I had been very bored with life. There had not seemed to be anything worth doing. But shortly after I came out of hospital, I dreamed that I was going to be executed. I suddenly realized that there were a lot of worthwhile things I could do if I were reprieved.

One result of my illness: when one is faced with the possibility of an early death, it makes one realize that life is worth living.

DEREK POWNEY

I was sitting there one evening with him when he asked, "Have you ever read John Donne's elegies?" Well, I think that Donne's elegies are the most beautiful love poetry I've ever read in my life, and also incredibly explicit—if they hadn't been written by John Donne they would be labeled pornographic without any doubt at all. So when Stephen launched into this I remember thinking, "Why on earth are you doing this, dear boy?" I couldn't understand why he'd suddenly got into this interest in Donne's poetry. Well, at about that time he met his wife, Jane, although we had not known this.

Stephen Hawking met Jane Wilde, who was just completing high school, at a New Year's party in St. Albans in January 1963, just before he entered the hospital for tests. Jane began language studies in London the following autumn.

STEPHEN HAWKING

*T*here did not seem much point in working at my research because I didn't expect to live long enough to finish my

53

Jane Wilde and Stephen Hawking at their wedding, 1965. His parents are standing at his right, hers at her left.

Ph.D. However, as time went by, the disease seemed to slow down. I began to understand general relativity and make progress with my work. But what really made a difference was that I got engaged to a woman named Jane Wilde. This gave me something to live for, but it also meant that I had to get a job if we were to get married.

⸻

In 1965 Stephen Hawking applied for, and received, a research fellowship at Caius College, Cambridge. He married Jane Wilde in July of the same year. Their first child, Robert, was born in 1967. A daughter, Lucy, was born in 1970, and a second son, Timothy, in 1979.

ISOBEL HAWKING

Stephen was already ill. Jane knew it, and it was another instance of Stephen's luck: meeting the right person at the right time. Stephen was very badly depressed and he wasn't really very much inclined to go on with work, having been told he only had two and a half years. But meeting Jane really put him on his mettle. And he started to work.

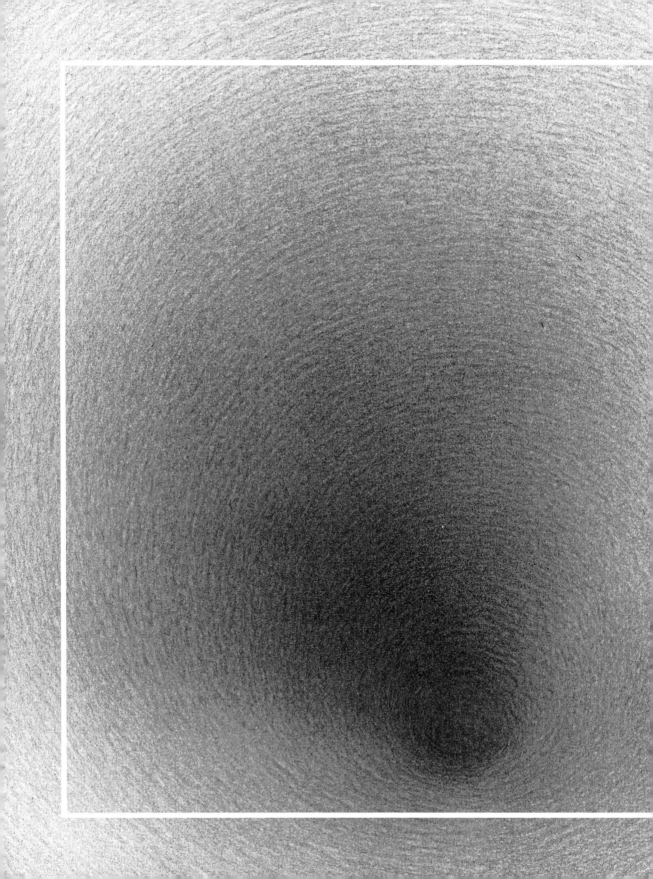

THREE

Before Stephen Hawking entered Cambridge in 1962, he considered two areas of theoretical physics for his research. One was cosmology, the study of the very large. The other was elementary particles, the study of the very small. However, he says, "I thought that elementary particles were less attractive, because there was no proper theory. All they could do was arrange the particles in families, as in botany. In cosmology, on the other hand, there was a well-defined theory, Einstein's general theory of relativity. At that time there was no one in Oxford in cosmology. However, at Cambridge there was Fred Hoyle, the most distinguished British astronomer of the time."

* * *

Albert Einstein developed two theories of relativity. The first, the special theory of relativity (1905), stated that light always travels at a constant speed, and that the speed of light is an absolute constant; all other motion is relative. In 1916 Einstein published a paper on his theory of general relativity. General relativity is essentially a theory of gravitation as the result of distortions in space-time geometry.

Ordinary geometry involves the distances between points and the angles between lines. However, on a curved surface, like the surface of the earth, these distances and angles don't obey the same geometrical laws that they do on a flat surface. For example, if two men start walking in different directions on a flat surface, they will get farther and farther apart. But if two men start walking in different directions on the surface of the earth, they will get farther and farther apart at first, but they will eventually meet up again on the other side of the earth.

Space-time geometry also involves distances and angles, but now one considers events, that is, points that are separated not only in space, but in time as well. The most important question is whether one can get from one event to another at the speed of light or slower.

Because gravity is a result of distortions in space-time geometry, gravitational fields affect measurements of time and distance. For example, general relativity predicts that an oscillating atom in the basement of a building should oscillate more slowly than an identical atom on the top floor. This effect is very small, but it has been measured (in a four-story building!) and the measurements agree with the prediction. Similar effects (but of much larger magnitude) are predicted to occur in very strong gravitational fields, like those that exist near a black hole.

FRED HOYLE

$\mathcal{T}$he word wisely used in regard to cosmology is "scenario." Science has two parts to it. It has a very accurate part that you get in theories such as quantum mechanics—this is exceedingly accurate, and anyone who challenges that is off his head.

But it has other parts in geology, astronomy, cosmology, biology, where theories are not really proved. They depend to a large extent for their acceptance on people making judgments. And it's a known phenomenon that where judgments are in-

Sir Fred Hoyle, educated at Cambridge, was Plumian Professor of Astronomy there and helped establish its Institute of Theoretical Astronomy in 1967, serving as its first director. Besides having written many works of science, he is also a prolific science fiction novelist.

volved, there's a strong tendency for people to pull together; that if you start with half the people making one particular judgment, they pull in the other half very quickly. It's a kind of herd instinct; I think it probably dates from the days when man was a hunting animal, and the worst thing you could do if you were in a community of, say, twenty men was to disagree about the direction in which you should hunt for the animal; it was better to choose one direction at random and all go in that direction than to split up and each go in a different direction—they needed the whole party to be successful.

We're not only controlled in the way we think by a few charismatic figures, but we're also very much controlled by what we can do. We tend to avoid the things that are too difficult for us; if we can solve certain equations, we tend to go that way. But the truth may lie in the difficult way. There's no guarantee that the universe is constructed explicitly to suit our standards of intelligence.

I think the word "scenario" is a very well chosen word. And I don't think that people in fifty years' time will hold anything like the views that we hold today; things will change quite a lot. And for that reason I prefer to tackle problems which in a sense are startling but which I can see are soluble.

* * *

With scientists Hermann Bondi and Thomas Gold, Fred Hoyle was one of the developers of steady state cosmology. The steady state theorists proposed that as the universe expands, and galaxies move apart from one another, matter is created from nothing to fill the void in space. This matter later condenses to form new stars and galaxies. Young newborn galaxies replace old dying ones, and the universe at any time looks much the same as at any other time. Thus, the universe is in a steady state.

The leading alternative theory is called big bang cosmology. The big bang cosmologists reject the creation of matter from nothing and instead argue that, since galaxies are now moving apart from one another, they must have been closer together in the past. Thus the universe in the very distant past must have been quite different from what it is today. Indeed, if one uses the equations of general relativity to trace the motions of the galaxies backward in time, one would find that the density of matter and the gravitational field would have been infinite. This point would have been the big bang.

Recent astronomical observations seem to strongly support big bang cosmology. They indicate that the universe in the past was very different from what it is now. As a result, the steady state theory has lost support. However, Hoyle believes that the evidence has been misinterpreted, and he continues to advocate the steady state theory.

STEPHEN HAWKING

*M*y application to do research at Cambridge was accepted, but to my annoyance, my supervisor was not Hoyle but a man called Dennis Sciama, of whom I had not heard. Sciama, like Hoyle, believed in the steady state theory, according to which the universe had no beginning or end in time.

In the end, however, this turned out to be for the best. Hoyle was abroad a lot, and I probably wouldn't have seen much of him. On the other hand, Sciama was there, and he was always stimulating, even though I often didn't agree with his ideas.

DENNIS SCIAMA

Dennis Sciama was a Lecturer in Mathematics in Cambridge from 1963 to 1970, was a Professor of Astrophysics at All Souls College in Oxford from 1970 to 1985, and now works at the International School of Advanced Studies in Trieste, Italy. He was Stephen Hawking's supervisor at Cambridge.

*T*here was at that time a somewhat acrimonious debate between some of the proponents of the steady state theory and observers who were testing it and, I think, hoping to disprove it. I played a very minor part at that time because I was a supporter of the steady state theory, not in the sense that I believed that it had to be true, but in that I found it so attractive I wanted it to be true.

When hostile observational evidence began to come in, Fred Hoyle took a leading part in trying to counter this evidence, and I played a small part at the side, also making suggestions as to how the hostile evidence could be answered. But as that evidence piled up, it became more and more evident that the game was up, and that one had to abandon the steady state theory.

The key year was probably 1965, not only because of the microwave background, but also because of the counts of radio sources which Martin Ryle, the leading radio astronomer at Cambridge, had been most prominent in pursuing. By a certain point he made even adherents like myself give up.

When I started research in relativity, there were only a handful of people in the world doing relativity in a serious

63

way. And then, in the early 1960s, the subject exploded in our faces. It was partly due to the fact that relativity was becoming much more exciting, and partly due to new developments in astronomy.

The very first development was around 1952, when people detected radio noise both from point sources called radio stars and from diffuse regions in our galaxy. What produced this radio noise was electrons moving in the magnetic field of the various galaxies. But the point about those electrons is that they were moving practically at the speed of light: they were cosmic ray electrons moving, as we say, relativistically—so the idea was that relativistic electrons, something previously studied mostly by relativists remote from astronomy, were responsible for the whole range of radioastronomy phenomena, which were themselves detected by great big pieces of metal, as it were.

The link between the abstract concepts of relativity and the very concrete methods of observing these radio emissions was very exciting. That was probably the point at which modern concepts in physics entered into astronomy in an observational way.

Quasars, or quasi-stellar objects, are starlike celestial bodies that emit enormous amounts of radiation. Discovered in 1963, they are thought to have formed at the beginning of the universe, fifteen billion years ago.

Pulsars are thought to be rotating neutron stars. The magnetic north and south of the neutron star is not lined up with the rotation, and this causes the pulses of radio waves. Pulsars were discovered in 1967.

After that first step we couldn't stop the progress. Almost every year an exciting new discovery was made, bringing the most exotic properties of modern physics into direct astronomical observation. There was the discovery of quasars and pulsars and the heat radiation that came from the big bang origin of the universe; and because quasars were thought to be related to gravitational collapse, even general relativity became important.

You see, special relativity was the first of

the two relativity theories of Einstein, the part of the theory which every physicist learns. But his theory of gravitation, the general theory, which is much more complicated and abstruse, was in the old days a special preserve of very professional experts. But when it was thought that gravitational collapse was important in explaining quasars, general relativity suddenly became important in astrophysics. And of course, if you could study the big bang origin of the universe with the radiation background, then cosmology is understandable only with general relativity. It's this development of the marriage of the abstract and the concrete that created this explosion of interest in relativity.

So naturally, when I set up a research school in Cambridge in the 1960s with students who seemed gifted enough to work in these difficult areas, these were the areas that I suggested they work in.

⋯⋯ ━━◆━━ ⋯⋯

No star lives forever; at some point, stars must burn off their fuel. Many will turn into white dwarfs—stable, small stars with a radius of a few thousand miles and a density of hundreds of tons per cubic inch. Other stars continue to collapse until they become neutron stars, much smaller than white dwarfs, with a radius of only ten miles and a density of hundreds of millions of tons per cubic inch. It is believed that other stars, above a certain size limit, collapse into what is called a black hole. A black hole is surrounded by a sphere called the event horizon. The event horizon is a one-way membrane. Although it is possible to pass from outside the event horizon into the black hole, nothing, including light, can travel in the opposite direction. Inside the event horizon, at the

A rendition of a black hole in quantum space: where light cannot escape, where time ends—and yet black holes are not as black as once thought.

center of the black hole, is a singularity, a place where the gravitational field is infinitely strong. Anyone who goes through the event horizon must eventually hit the singularity, with disastrous results.

Kip Thorne received his B.S. from the California Institute of Technology in 1962 and his Ph.D. in physics under John Wheeler at Princeton in 1965. He is currently the William R. Kenan, Jr., Professor, and Professor of Theoretical Physics, at Caltech.

KIP THORNE

The theory of the gravitational collapse of stars and black holes began in the late 1930s. Robert Oppenheimer and his students, who really started the subject, took off from the earlier work of Lev Landau, who was the father of modern theoretical physics in the Soviet Union.

Landau had puzzled over how stars get the energy that keeps them hot; he had devised the idea that, at the center of stars like the sun, there might be a neutron star with a size of, say, ten or twenty kilometers, and a mass of some small fraction of the mass of the sun, say a tenth of the mass of the sun; and that the gas of the sun might be gradually sinking onto this neutron star. And this sinking would produce the heat that keeps the sun hot. He massaged this idea in his own head and in discussions with colleagues during the 1930s.

But there came a time when he felt the fire of Stalin's purges breathing down his neck. At that point he searched around desperately for something that would make a big

68

splash with the public press, something that would protect him from Stalin's purges—Landau was suspect because he'd spent a lot of time in Germany in the early 1930s, during the period when he was learning physics. Because of that he had been accused by some Soviet physicists of spying for Germany, despite the fact that he was Jewish.

So Landau sent Niels Bohr in Copenhagen a manuscript containing this idea of keeping the sun hot by a neutron core inside it, along with a letter requesting that, if Bohr thought this was a good idea, he transmit it to *Nature*. Bohr did so, and then immediately thereafter he received a telegram from *Izvestia* asking Bohr what he thought of this work. Bohr sent back a glowing report, which *Izvestia* immediately published.

Niels Bohr (1885-1962) was a Danish physicist who won a Nobel Prize in 1922. He was the leading figure in the developments in atomic theory that led to the invention of quantum mechanics.

Bohr knew this was Landau's attempt to keep out of prison. But it didn't work. Landau was thrown into prison for a year and nearly died.

During the period Landau was imprisoned, Oppenheimer and his student Bob Serber looked at the Landau theory and saw a hole in it. They figured, "Well, Landau, he can take the heat, he's a great physicist," so they published a paper in the *Physical Review* tearing Landau's idea apart while Landau was in prison.

So what got Oppenheimer and his student started thinking about neutron stars, and then black holes, was Landau's work, which they didn't realize was an appeal to keep himself out of prison.

Landau did emerge from prison—a result of great pressure from Soviet scientists—about a year after he went in. He was in very poor health. But I don't think it was known in the West until very recently that this was in fact the origin of Oppenheimer's efforts, with his student

Hartland Snyder, who had been a truck driver in Utah before he became Oppenheimer's student, that led to the theory of the black hole.

Oppenheimer and Snyder's first step, having discovered that the sun and other stars could not be kept hot by a neutron core, was to ask themselves: Well, suppose that you have neutron stars that are the remnant of the death of a normal kind of a star—how big can they be? Oppenheimer and another student, George Volkoff, showed that there is a maximum mass for a neutron star, that neutron stars cannot get any larger than, well, their estimate was 0.7 solar masses. We now know that it's more like two solar masses, because we now understand nuclear physics better.

But having seen that there's a maximum possible mass for a neutron star, Oppenheimer then took the next step, asking himself what happens when a massive star dies. Using general relativity, Oppenheimer and Snyder computed the implosion of a star, and they saw that the star would cut itself off from the exterior universe—would, in the fancy words we use nowadays, "go inside its own horizon."

They refused, however, to consider the issue of what happens to the star inside the horizon; they simply saw in the equations that the star cuts itself off from the rest of the universe. Oppenheimer was not a speculative person. He could look at a very complicated physical problem, identify the key things going on in it, solve the problem, and predict what would happen—the tools that were needed for designing the hydrogen bomb. But he refused even to ask general relativity what happens inside the horizon, the very questions that have become so interesting more recently with Stephen Hawking's work on quantum gravity.

In 1974 Stephen Hawking discovered that the event horizon was not strictly impermeable when quantum mechanics was taken into consideration. Black holes radiate energy and lose mass. The more massive a black hole, the less rapidly it loses its mass; the effect is very small for stellar-mass black holes. Nevertheless, all black holes eventually radiate away all their mass and disappear.

· · · · · · ——•—•———— · · · · · ·

Up to about 1910 it was thought that matter was made out of particles like billiard balls that had definite positions and speeds, and whose behavior could be predicted precisely by the laws of physics. However, evidence began to come from experiments that these precise so-called classical laws had to be replaced at very short distances by what were called quantum laws. According to these quantum laws, particles didn't have exactly defined positions or speeds but instead were smeared out with a probability distribution, or wave function, which measured the probability of finding the particle at different positions. The quantum laws implied that one couldn't measure both the position and the speed of a particle. The more accurately one measured the position, the less accurately one could measure the speed, and vice versa.

The novel features of general relativity are most striking in strong gravitational fields. Those of quantum mechanics are most striking at small distances. Thus a quantum-mechanical theory of space-time geometry, quantum gravity, should be essential to understanding events that take place at small distances and involve strong gravitational fields. One such event is the big bang. Another would be inside a black hole.

· · · · · · ——•—•———— · · · · · ·

Antony Hewish won a Nobel Prize in Physics in 1974, along with Sir Martin Ryle, for the discovery of pulsars, a discovery that proved the existence of neutron stars and that makes the phenomena of black holes more probable. He has been Professor of Radio Astronomy at Cambridge since 1971.

ANTONY HEWISH

When radio telescopes first began to pick up radio waves from the universe, it was done with very crude equipment. The first real excitement took place when the first radio-emitting galaxy was identified by people using the big optical telescope in California, at Palomar. These strange objects—we didn't know what they were at that time—were points in the sky emitting radio waves—but what *were* they? They weren't the sun or any known star.

It turned out that with the optical telescope you could just see a faint smudge, and we learned that this smudge was actually a galaxy of a kind never seen before, at a distance of around a thousand million light-years. So, with simple apparatus we had picked up a galaxy at an enormously great distance; we were looking at it way back in history, a thousand million years ago. And it was clear that with better equipment, we could detect objects much fainter than this with the radio telescopes, and therefore, presumably, further away.

This raised the whole possibility that you could check the competing cosmological theories by looking back in time, looking back over the history of the universe. Before this, cosmology had been the battle of theoreticians. Now it became what we could call an observational science, something we could really observe.

The curtain was drawn back on the universe; we now had a much deeper look into space.

It turned out that as you looked back in time, there were

far more of these radio galaxies than would fit the steady state theory of the universe of Hoyle, Bondi, and Gold. In the universe of the steady state theory, stars and galaxies are forming and are decaying, but as the universe expands—and everybody agrees the universe is expanding—material must be added into it to make new stars and galaxies, so that, on average, the picture of the universe looks the same. If the universe is in a steady state—a kind of balance, if you like—looking at it at one time or another makes no difference: on average, it should look the same. So if you can observe very distant objects, then you can look back in time and see if the universe was the same in the past as it is today, or whether it's different.

The first results coming from the radio telescopes suggested that we had a very different universe. It had far more of these radio galaxies in it than a smooth, steady state universe would have. So the universe wasn't in a state of continuous creation. It looked much more like an evolution in time.

The work that was done on radio galaxies seemed to point very definitely to a universe which had an evolutionary history. This was confirmed dramatically in 1965, when Arno Penzias and Robert Wilson in the United States picked up this cosmic background radiation on their radio telescope—residual heat radiation left over from the hot big bang which created the universe—confirming that the universe couldn't be in a steady state.

The faint heat radiation that their radio telescope picked up is very cool—it corresponds to a sky background just under three degrees Kelvin in temperature, which is very cold indeed. But if you work out in cosmological theory how that radiation originated, it tells you that once upon a time the universe was incredibly hot, with a temperature of millions of degrees; what we're picking up now is a sort of relic, a fossil radiation from the very earliest phase of the universe, which fits in with the idea of the big bang, the sudden creation—but which makes it a hot big bang.

This radiation we pick up, this fossil, says that long ago in the past there was a very hot, very condensed universe, and roughly speaking, you can imagine this as a kind of cosmic explosion which launched the universe on its career. Now we see it expanding and cooling as time goes on.

DENNIS SCIAMA

I remember the very first seminar that was given about pulsars in Cambridge. The talk was given an apparently innocuous title—it was just called, I think, "A New Class of Radio Sources." Tony Hewish was going to give it. But a rumor had got around that it was not just any old new class of radio sources but something spectacular, so the meeting had to be moved from the normal seminar room of the radio astronomers to a very large lecture room, and the place was packed. The rumor had spread well.

And at that meeting pulsars were first announced. There was some discussion about what they were; they clearly had to be very compact objects, but it wasn't certain whether they were white dwarfs, which are compact objects that, though very exotic, were already well known to astronomers, or whether they were the so-called neutron stars, which would be very much more compact than white dwarfs and almost in the black-hole state. It took a few months of discussion before it became clear. Tommy Gold, who had worked earlier with Hoyle and Bondi in Cambridge on the steady state theory, was the first to make the clear argument that pulsars were rotating neutron stars and couldn't be anything else.

So, for the first time, these very compact objects that had before been purely theoretical constructs, that had previously not been taken seriously by working astronomers, suddenly became the objects at the center of a class of radio sources that radio astronomers could observe all over the world. And since neutron stars are almost in the black-hole condition—the radius of a neutron star is only perhaps a few times larger than the radius that a black hole of that mass would have—it gave confidence to people who took the concept of black holes seriously.

ANTONY HEWISH

*B*ack in the 1930s, when James Chadwick discovered the neutron particle, it was worked out that you could have very strange objects—gravity is an immensely powerful force, and if you have a star which is running short of its nuclear fuel, gravity is going to condense that star, even if it is the size of the sun, into a ball a few miles across, turning most of the star's material into these neutron particles. It's like crushing matter as we know it out of existence, squeezing positive and negative charges so close together that they fuse to become a new kind of particle.

So there was this idea that there might be neutron stars, and it was great good fortune that the experimental program I was following in Cambridge really led directly to the discovery of these things. You see, I'd set up this experiment which was actually designed to observe quasars. I had discovered that if you look at certain radio galaxies through the sun's atmosphere, they would kind of, well, twinkle, like a star. But they only do that if they have the incredibly compact dimensions which are typical of quasars. Quasars are immensely powerful galaxies, but their energy source comes from a tiny volume in the middle of them—because they're such highly compact objects, they're very small even when seen with a radio telescope. So it is from this twinkling that you can tell whether you're looking at a normal kind of extended radio galaxy, or whether you're looking at something which has this compact structure in it.

So I designed this radio telescope which was unlike anything that had been seen before in radio astronomy; it was created to look at this twinkling effect, which meant it had properties quite unlike anything else. It worked at a long wavelength; we were looking repeatedly at the sky and so on for fluctuating sources. Of course, it wasn't long before we picked up the pulses. It was a great stroke of good luck that the parameters of the telescope exactly fitted what you needed to pick up pulses.

These pulsars turned out to be rotating neutron stars; and this discovery was made in 1967 and became widely known in 1968. We didn't know what they were then; but they had to be tiny because nothing else would fit the bill. When the first publication came out, I suggested vibrating neutron stars, or vibrating white dwarf stars, but the neutron star looked a bit more likely. Twelve months later this idea was confirmed, and neutron stars entered astrophysics.

It was all very exciting. I mean, who would dream that you would get intelligent-looking signals coming from the sky? What on earth in the sky could be launching pulses? We considered all kinds of things and then checked them out, such as American spacecraft that nobody had told us about, or signals reflected in from the moon. But when these local possibilities had been removed, I certainly and seriously began to think that we might be picking up for the first time real, intelligent signals—little green men as we called them—the signals from an alien civilization somewhere.

However, my research student Jocelyn Bell had been looking through further records, and we came up with more and more of these pulsing signals until it became clear that we had to look for another explanation; it wasn't little green men—although for a while I did take that quite seriously; you couldn't dismiss it easily.

It's like a detective story: if there's only one solution, that's it, there's only one person who can have done the crime. Well, when the planetary motion had been removed, I knew that the pulsation couldn't be coming from a planet. The pulses were very narrow, and this meant that the emitter had to be very small; you can't have a large body emitting short, sharp pulses because of the travel time of the radiation from different parts of it. It had to be something highly compact; it had to be an object smaller than a few thousand kilometers in size, yet at the distance of a star.

STEPHEN HAWKING

I went to the seminar where pulsars were announced. The room was decorated with paper little green men. The first four pulsars discovered were called LGM one through four. "LGM" stood for "little green men."

KIP THORNE

I first met Stephen at an international conference on general relativity and gravitation. I'd just gotten my Ph.D., while Stephen was in the late stages of his Ph.D. research at Cambridge. At the time he was walking with a cane and he was a little wobbly. His speech was very clear, however, with just a bit of hesitation in it, and I didn't really understand until later what the meaning of his disease was.

This meeting came at the time when Stephen was in the early stages of his work on singularities in cosmology, in the large-scale structure of the universe, using techniques that had been pioneered by Roger Penrose. Our brief conversation in a tearoom was enough to leave me very impressed with the things he was doing, with the ideas he was developing, ideas clearly based on the fundamental techniques that Penrose had devised, but being carried off into the cosmological context, whereas Penrose had used them in the context of black holes.

The sense in which Stephen was perhaps unique or powerful was that he got into all this far faster than other people—he mastered the techniques, started using them, and simply took off so fast that he left everybody else in the dust. In retrospect, I think it was the obvious thing to do, but one always wonders how much one's retrospective view represents the real situation.

ROGER PENROSE

I remember having a very animated conversation with my friend Ivor Robinson, and then we had to cross a road. As we crossed, of course, the conversation stopped. Then we got to the other side, and evidently I had had some idea in crossing the road, but then the conversation started up again and it got completely blotted out of my mind. It was only later, after my friend had gone home, that I began to have this strange feeling of elation, of feeling wonderful. And I couldn't figure out why on earth I should feel like that. So I went back over the day, thinking of all the possible things which might have contributed to such a feeling. Then gradually I unearthed this thought which I'd had while crossing the road.

· · · · · · ———◆——— · · · · · ·

Penrose's idea was a way to show that if a star collapses beyond a certain point, it couldn't re-expand. Instead, a singularity of space-time would occur, a point at which time came to an end and the laws of physics broke down. Before Penrose's result, it was thought that if a star was not perfectly round or if it was rotating a bit, the matter in the collapsing star might avoid becoming infinitely dense. Instead, the collapsing matter might whiz past itself and expand again.

· · · · · · ———◆——— · · · · · ·

DENNIS SCIAMA

*P*enrose announced his result: that when stars collapse indefinitely, they will become singular, as long as some very broad conditions, conditions that everybody would have regarded as reasonable, are satisfied.

I remember Stephen Hawking, who was then approaching his third year as a research student, saying, "What very

79

interesting results—I wonder whether they could be adapted to understanding the origin of the universe?" What he had in mind, you see, was that if you reverse the sense of time—just mentally—you can think of the expanding universe as a collapsing system; it's a bit like a very giant star collapsing. And the same considerations apply: as the universe collapses, in that sense of time, it reaches a singularity; or, in the normal sense of time, we have the explosion from the big bang, which is certainly singular in the very symmetric models of the universe.

Then Stephen said, "Maybe the same consideration applies as in Penrose's theorem for stars." That is, even in a realistically irregular universe, maybe the beginning had to be singular, which again would not just be an interesting result; it would create an intellectual crisis because it would mean that general relativity breaks down at the very beginning of the universe. So Stephen said, "I'm going to try and adapt Penrose's result to the whole universe."

In his last year as a research student he did just that. It was a nontrivial matter to adapt Penrose's method: it was a brilliant piece of work. And if you look at his thesis, you will find that the last chapter contains his first singularity theorem for the beginning of the universe.

STEPHEN HAWKING

Penrose's result was the first singularity theorem. To prove it, Penrose introduced some completely new techniques into general relativity. I was not at the seminar in London at which he presented his theorem, but I heard about it the next day. I was excited because I had been considering the rather similar problem of whether there had to be a singularity in the past which was a beginning of time, or whether there could have been a previous contracting phase of the universe and a bounce. I was able to use Penrose's techniques and add some of my own to show that if classical general relativity was correct, there must have been a singularity in the past, which was a beginning of time. Anything that might exist before the singularity could not be considered part of the universe.

Bernard Carr did his graduate work with Stephen Hawking; he also accompanied the Hawking family to the California Institute of Technology in 1974, where he helped look after Stephen. Today he teaches physics at Queen Mary and Westfield College, London University.

BERNARD CARR

Stephen and Roger Penrose are the two great relativists, and they worked on very similar problems. Eventually they came up with the famous theorems which say that at the beginning of the universe there must have been a singularity, or a point where relativity theory breaks down. In a sense, what they discovered was that relativity theory, although it's very beautiful as a description of the universe, actually predicts its own downfall under very special circumstances.

Singularities arise in two contexts: they arise at the center of black holes, which is something Roger Penrose proved; but we can also show that there must have been a singularity in the very early universe, and that was something that Hawking and Penrose proved together.

If classical relativity breaks down at

singularities, then the question is, what does happen? One could just say, well, we give up because we know the theory breaks down, but of course, the goal of quantum gravity is to find out.

STEPHEN HAWKING

$\mathcal{O}$bservations of distant galaxies indicate they are moving away from us: the universe is expanding. This implies that the galaxies must have been closer together in the past. The question then arises: Was there a time in the past when all the galaxies were on top of each other and the density of the universe was infinite? Or was there a previous contracting phase, in which the galaxies managed to avoid hitting each other? Maybe they flew past each other and started to move away from each other. To answer this question required new mathematical techniques. These were developed between 1965 and 1970, mainly by Roger Penrose and myself. We used these techniques to show that there must have been a state of infinite density in the past, if the general theory of relativity was correct.

This state of infinite density is called the big bang singularity. It would be the beginning of the universe. All the known laws of science would break down at a singularity. This would mean that science would not be able to predict how the universe would begin, if general relativity is correct. However, my more recent work indicates that it is possible to predict how the universe would begin if one takes into account the theory of quantum mechanics, the theory of the very small.

Kip Thorne

In the 1950s and earlier, general relativity was largely a branch of mathematics; it was chiefly John Wheeler at Princeton who brought it into physics. A physicist wants to ask questions about the real world, questions such as: What is the nature of elementary particles? What lies under elementary particles? Whereas the people who had dominated general relativity for the preceding few decades were typically asking questions about whether they could derive in a mathematically rigorous way the laws that govern the motions of idealized particles in curved space-time— questions motivated largely by searches for deep mathematical understanding of the structure of the theory, but not questions which were trying to get to the heart of the nature of the universe.

So the 1950s were a period of a major change in the way people thought about curved space-time, moving from a situation where people looked at the structure of the mathematics to where people began to think in terms of physics. This was done thanks to Wheeler: some people have called him the Pied Piper of black holes because of his enthusiastic, compelling vision that this was the most exciting issue in physics, the place where one should be if one really wants to understand the universe and the fundamental laws of physics.

Johnnie gave lectures at Cambridge at roughly the same period which strongly influenced people in Dennis Sciama's group and, ultimately, Stephen Hawking. He has tremendously good judgment about what's important in physics, and his influence in convincing us that this is where the action is was terribly important to the history of the subject.

John Wheeler, editor, coauthor, or author of eight books, is a professor emeritus at both the University of Texas at Austin and Princeton University, the recipient of honorary degrees from sixteen universities, and the physicist who in 1969 coined the term "black hole."

JOHN WHEELER

In 1969 we had this meeting at the Institute of Space Physics on Amsterdam Avenue in New York City. I said that before we reached a final conclusion we ought to throw into the pot still another object, a gravitationally completely collapsed object. Well, after you've used a phrase like a "gravitationally completely collapsed object" ten times, you conclude you've got to get a better name. So that's when I switched to the term "black hole."

BRANDON CARTER

It was Oppenheimer who first started thinking of the implications and the formulations of what twenty years later was called a black hole. But it wasn't until the 1950s that people started to take up this lead. John Wheeler was particularly instrumental in that. And on the English side, a key personality was my own supervisor, also Stephen Hawking's supervisor, Dennis Sciama. Dennis played a role in England rather analogous to that of Wheeler: he got people interested. One of

Brandon Carter was born in Australia and educated in Scotland before completing his undergraduate work at Pembroke College, Cambridge. He continued to study and teach in the Department of Applied Mathematics and Theoretical Physics (DAMTP) at Cambridge until 1975, when he went to work for France's Centre National de la Recherche Scientifique. Since 1986 he has been the Director of Research at the Observatory of Paris.

those people was Roger Penrose. Penrose's unique role was to use a lot of modern mathematical methods, advanced topology and differential geometry, to solve problems which a pure mathematician would never have thought of tackling.

It made a big difference when the term "black hole" was invented. As long as there was no generally accepted word, it was not only difficult to communicate to people in other fields what we were talking about, it was difficult even with specialists. There could be groups of people effectively working on the same problems but not quite knowing this, because of a lack of a word.

Things changed dramatically when John Wheeler invented the term. This wasn't the first attempt; other terms had been used, but they hadn't caught on. The magic is when something does catch on. Everybody adopted it, and from then on, people around the world, in Moscow, in America, in England, and elsewhere, could know they were speaking about the same thing. And not only that: suddenly the whole range of concepts got through to the general public—even science fiction writers could talk about it.

JOHN WHEELER

*F*riends ask me, "What if a black hole is black? How can you see it?" And I say, have you ever been to a ball? Have you ever watched the young men dressed in their black evening tuxedos and the girls in their white dresses whirling around, held in each other's arms, and the lights turned low? And all you can see is the girls. Well, the girl is the ordinary star and the boy is the black hole. You can't see the black hole any more than you can see the boy. But the girl going around gives you convincing evidence that there must be something there holding her in orbit.

STEPHEN HAWKING

Falling into a black hole has become one of the horrors of science fiction. In fact, black holes can now be said to be really matters of science fact.

Of course, where the science fiction writers really go to town is on what happens if you do fall into a black hole. A common suggestion is that if the black hole is rotating, you can fall through a little hole in space-time and out into another region of the universe. This obviously raises great possibilities for space travel. Indeed, we need something like this if travel to other stars, let alone other galaxies, is to be a practical proposition in the future. Otherwise, the fact that nothing can travel faster than light means that the round trip to the nearest star would take at least eight years. So much for a weekend break on Alpha Centauri. On the other hand, if one could pass through a black hole, one might re-emerge anywhere in the universe. Quite how you choose your destination is not clear: you might set out for a holiday in Virgo and end up in the Crab Nebula.

I'm sorry to disappoint prospective galactic tourists, but this scenario doesn't work: if you jump into a black hole, you will get torn apart and crushed out of existence. However, there is a sense in which the particles that make up your body do carry on into another universe. I don't know if it would be much consolation to someone being made into spaghetti in a black hole to know that his particles might survive.

JOHN TAYLOR

John Taylor is Professor of Mathematics at King's College, London. He did much research on black holes in the 1970s and wrote a best-selling book on the subject. He has recently moved away from the study of cosmology and is now working on neural systems.

As one falls into a black hole, if the black hole is very big—say it's been made from matter the size of our galaxy, so that it has an event horizon about the radius of the solar system—one can fall through the event horizon without feeling anything, without noticing it. Then, after about a week of falling, one begins to feel the pinch, and one extends longer and longer and gets slightly thinner; and, of course, one begins to get squeezed, until one gets very long and very thin and rather nasty. By the end of two weeks, one's fallen right into the center, and is dead. That, of course, is the problem with gravitational collapse.

But what happens at the center? At the center, standard gravity, with a little bit of classical matter added in, would tell you that you disappear. Now, that's absurd. It's absolutely terrible; you're destroying the whole structure of the model upon which you make these predictions.

I was once asked to actually be an adjudicator on an essay for which the subject was how to fall through a black hole and live. The problem I had was that I wouldn't know how to give out the prize. If I said, "Well, that looks like a good essay," the only real way of showing that this was right was to follow the experiment and fall into a black hole. But then, having fallen in—and I assume you would be taking the person who wrote the essay in with you—the question would be: How do you tell the rest of the world? Do you take the prize in with you that you give to them? And what do they do with it when they get to the center?

Time, inside a black hole. The watch never reaches midnight, and no one ever reaches the center of the hole.

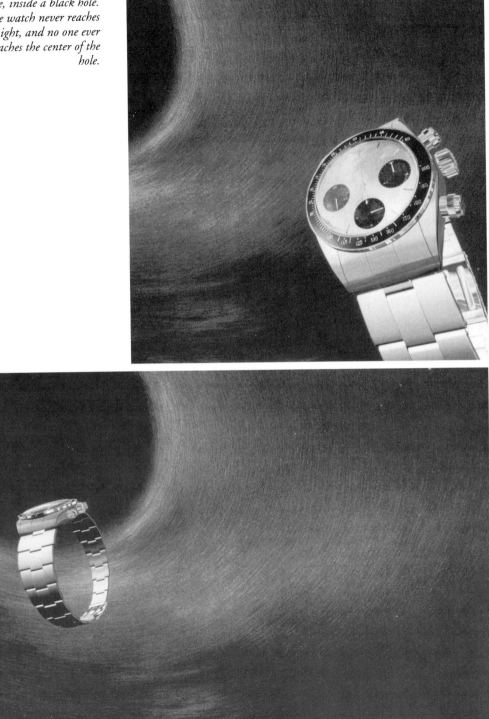

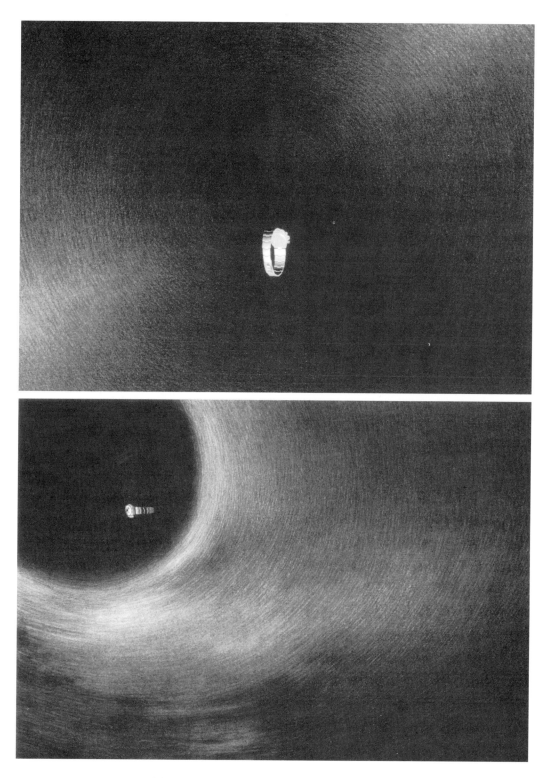

Actually, the model of gravity and classical matter contains in itself the seeds of its own destruction, by the fact that the matter falling into the center of a black hole would suddenly disappear. Now, when it gets very compressed, maybe in the last billion-billion-billion-billion-billionth of a second, at that point and from then on, the nature of time and space may have changed considerably. And so in order to probe that last, absolutely very brief time, one needs to use effects that come in at very high energies, because as all the matter is being channeled in and being collapsed, it heats up by compressing itself and gets to very high energies, and effectively you're going to rerun the beginning of the whole universe—you're going back to a very collapsed state.

And to describe that, we need these quantum effects, these uncertainty effects that may allow you to avoid actually disappearing. Now, there may be a number of reasons for how that occurs. One very nice one that might be right, and that might get us off the hook as far as the first cause is concerned, is that as you go back earlier and earlier into the beginning of the big bang, or as you fall inside a black hole to the center, time is going on, your wristwatch time—of course, your wristwatch is destroyed; so is your wrist, sadly enough—activities are going on: and time is activity. Now, as you go closer and closer to the center, it may be that the activity time stretches out, so it's not a billion-billion-billion-billion-billionth of a second that's got to pass; it stretches out, so you go back and back and back, or on and on and on, into the center of a black hole, and you never quite reach the center because there are always new activities occurring. Think of an onion with an infinite number of layers, and you peel off layer after layer after layer; you can only do each peeling at a finite time, and you're trying to get to the center of this onion. As you peel off each layer, that's a new activity, and you can do it, say, once a second. It may be like that, falling into a black hole. It most likely is like that, I feel, at the beginning of the universe. In which case, if there is activity always going back, there never was a beginning, and so time has stretched out and there could be no first cause. And the reason that we ask about a first cause is because we don't appreciate that time has changed so enormously in that very strange, very bizarre environment; and it's that which we must take account of properly.

As one falls through a black hole, one can look backwards, of course, and see other things falling in as well, as you would expect. One will see some very bizarre distortions indeed as one falls through. If one avoids the singular ring, one could fall through into a further universe. That further universe should be a copy of our own to a certain extent, but whether the details of its evolution are the same as our own, we don't know.

One could conjecture that, for example, one might not meet one's own star, one's own planet, one's kind there at all: there may be completely different environments altogether. Certainly it boggles the mind completely as to what could happen in these alternate universes.

Brandon Carter

*B*efore you lose sight of the outer world, you would see things happening, and you would see them happening at such a great rate that it would look like a fireworks display. The frustration would be that although you would be able to see everything that happens in the future, it would be going so fast that, from a scientific point of view, you would have no time to analyze it. You wouldn't be able to take it

91

in, and eventually things would be going off so fast and it would be so explosive that you yourself would be destroyed by the explosion. It would be a very exciting way to end one's life. It would be the way I would choose if I had the choice.

STEPHEN HAWKING

One evening shortly after the birth of my daughter, Lucy, I started to think about black holes as I was getting into bed. My disability made this rather a slow process, so I had plenty of time. Suddenly I realized that the area of the event horizon always increases with time. I was so excited with my discovery that I didn't get much sleep that night. The increase in the area of the event horizon suggested that a black hole had a quantity called entropy, which measured the amount of disorder it contained; and if it had an entropy, it must have a temperature. However, if you heat up a poker in the fire, it glows red-hot and emits radiation. But a black hole cannot emit radiation, because nothing can escape from a black hole.

DENNIS SCIAMA

Stephen Hawking's famous paper on the radiating black holes dates from 1974. What had happened was that an Israeli physicist named Jacob Bekenstein had suggested a year or two earlier that certain curious properties of black holes made them look as though they were thermodynamic systems. Bekenstein had suggested that perhaps a black hole had a temperature and an entropy

Jacob D. Bekenstein was a doctoral student of John Wheeler's at Princeton University. He taught for many years at Ben Gurion University in Beersheba, Israel, and is now at the Racah Institute of Physics of the Hebrew University in Jerusalem.

just like an ordinary hot body, with the temperature being inversely proportional to the mass of the black hole. That would mean that a low-mass black hole would be hotter than a heavy black hole, and that the entropy would be proportional to the area of the horizon of the black hole.

...... ———•——— ......

Thermodynamics is the division of physics that concerns the relationship between heat and energy. The second law of thermodynamics is perhaps the most famous. It says that the entropy, or the disorder, of an isolated system always increases: once an egg falls to the floor and breaks, it is not likely ever to re-form into its original shape.

...... ———•——— ......

STEPHEN HAWKING

General relativity is what is called a classical theory. It predicts a single definite path for each particle. But according to quantum mechanics, the other great theory of the twentieth century, there is an element of chance and uncertainty. While I was visiting Moscow in 1973, I discussed the effect of quantum mechanics on black holes with Yakov Zeldovich, the father of the Soviet hydrogen bomb. Shortly afterwards I made my most surprising discovery. I found that particles would leak through the event horizon and escape from the black hole.

Sciama was the first person I told, but I soon realized that the cat was out of the bag. Roger Penrose phoned up while

I was eating a birthday dinner. He was very excited and he went on so long that my dinner was quite cold. It was a great pity, because it was goose, which I'm very fond of.

JOHN WHEELER

Jacob Bekenstein came into the office one day. "Jacob," I said, "it always troubles me that when I put a hot teacup next to a cold teacup, I've increased the amount of disorder in the universe by letting the heat flow from one to the other, and I've committed a crime which will echo down the corridors of time to all eternity. But Jacob, if a black hole swims by, and I drop both teacups into this, I've concealed the evidence of my crime, have I not?"

Jacob looked troubled and he came back to me later and said, "No, you have not concealed the evidence of your crime. The black hole records what's happened to you. The black hole has increased its entropy, its degree of disorder, and that's where the evidence of your crime will show up forever."

Bekenstein gave an approximate figure for how much disorder is in a black hole. Stephen Hawking and Brandon Carter read Bekenstein's result, and were so upset by it that they set out to prove it was wrong. But as they studied it, they found they came to agree with his conclusion. Stephen Hawking went even further and devised a scheme to understand how particles would come out of the black hole; and its temperature was not just a theoretical temperature, it's a temperature at which particles would really evaporate from the black hole.

A black hole is not totally black. A black hole as massive as the sun can have a temperature of one millionth of a degree above absolute zero—peanuts, but still something.

STEPHEN HAWKING

In a paper I wrote with Brandon Carter, we pointed out a fatal flaw in Bekenstein's idea. If black holes have an entropy, they ought to have a temperature. And if they have a temperature, they ought to give off radiation. But how could they give off radiation if nothing can escape from a black hole? I must admit that in writing this paper I was motivated partly by irritation with Bekenstein, who I felt had misused my discovery of the increase of the area of the event horizon. It turned out in the end that he was basically correct, though in a manner he had certainly not expected.

DENNIS SCIAMA

At first Hawking didn't believe these analogies to thermodynamics were more than just curious. Bekenstein argued that they were really thermodynamic properties. But you can follow the evolution of Stephen Hawking's thought, because in 1973 he wrote a very important paper, with Jim Bardeen and Brandon Carter, which they very pointedly called "The Four Laws of Black Hole Mechanics," not "The Four Laws of Black Hole Thermodynamics." They explored these analogies but emphasized that this wasn't real thermodynamics—and they had a good reason for saying that. Because at that time it was supposed that a hot black hole still could not radiate radiation; and if a hot body doesn't radiate, you've lost your thermodynamics, it just doesn't work. Therefore these analogies were not deep.

That was in 1973. Then, sometime between 1973 and 1974, Hawking was looking at what would happen when a star collapsed and you actually introduced quantum-

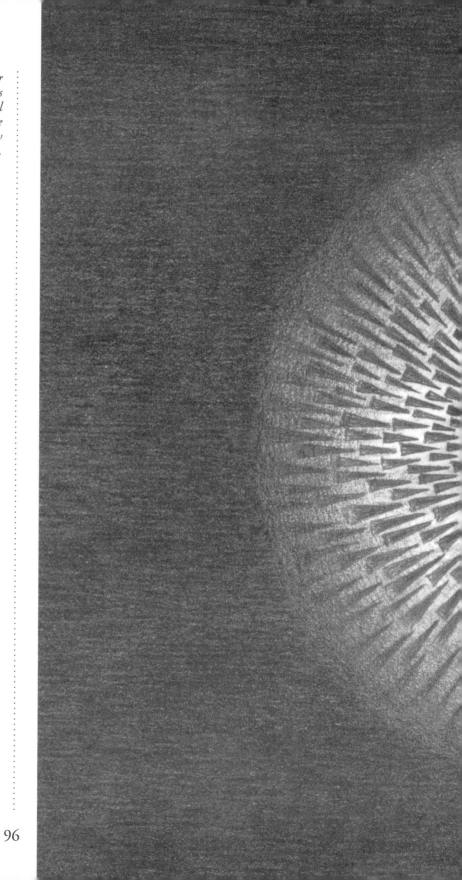

A rendition of stellar collapse. A star that has burned up most of its fuel will eventually condense into a ball only a few miles across.

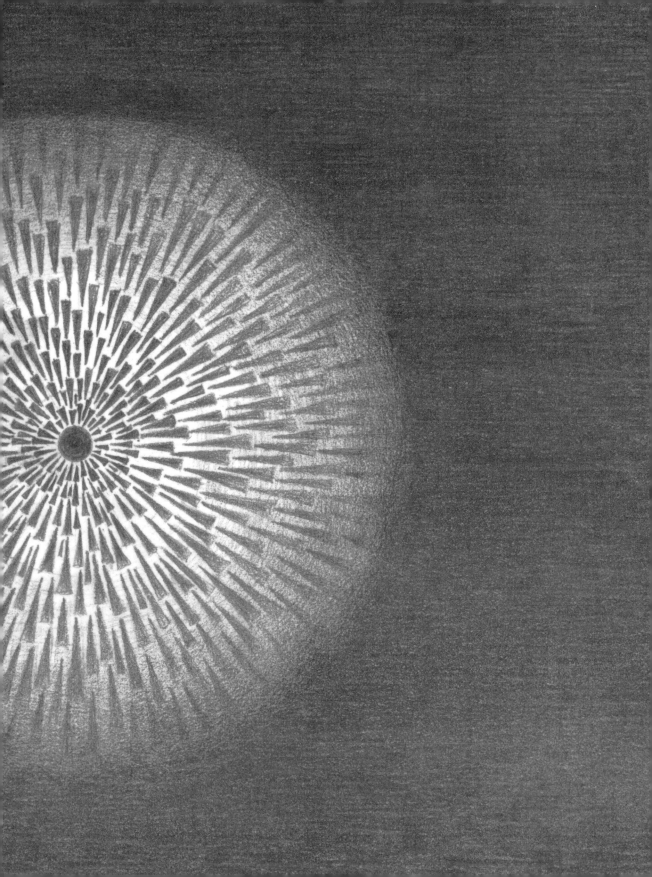

mechanical effects into the region outside the collapsing star. He discovered, by means of a complicated calculation—to me it's a miracle because it's so complicated and messy—that in the changing gravitational field of the collapsing star, there would be radiation.

Well, everybody would expect that, because when a field changes, then, in quantum mechanics, particles are created and radiated. What emerged from this calculation that was unexpected was that at the end of the collapse, when the star was approaching the black hole condition—that is, when it became so compact that radiation could no longer leave the black hole, at least classically—there was a residual radiation, and this residual radiation had a thermal spectrum characterized by the temperature of the black hole, as previously introduced by Bekenstein. That was the key discovery.

I remember that when I was visiting Cambridge in early 1974, I met Martin Rees, who was shaking with excitement. He said, "Have you heard? Have you heard what Stephen has discovered? Everything is different, everything has changed!"

Professor Martin Rees is a colleague of Stephen Hawking's who is currently working at the Institute of Astronomy in Cambridge.

"What are you talking about?" I asked. And Rees explained to me that black holes aren't black, because, with this quantum-mechanical effect, they radiate like hot bodies. This introduced a new unification of thermodynamics, general relativity, and quantum mechanics that would change our understanding of physics.

Later Stephen came to a meeting that I was helping to organize near Oxford, at the Rutherford–Appleton Laboratory, and people were flabbergasted. I remember someone getting up and saying, "You must be wrong, Stephen, I don't believe a word of it!"

JOHN TAYLOR

I once said that I was unhappy with the explanation given in terms of negative-energy particles being created. But I feel this is part of the controversy of science. You must have the give-and-take. And I'm delighted to be able to be part of that. That's what makes it fun. You know, if you all sat down and said, "Oh, lovely," whenever you do have niggling problems in your minds, that's not doing a service to science. But I was not antagonistic, except for that one time when I questioned him.

STEPHEN HAWKING

I still didn't completely believe it. I finally convinced myself that black holes radiate when I found a mechanism by which this could happen. According to quantum mechanics, space is filled with virtual particles and antiparticles that are constantly materializing in pairs, separating, and then coming together again and annihilating each other. In the presence of a black hole, one member of a pair of virtual particles may fall into the hole, leaving the other member without a partner with which to annihilate. The forsaken particle appears to be radiation emitted by the black holes. Quantum mechanics has allowed a particle to escape from a black hole, something that is not allowed by Einstein's general theory of relativity.

A virtual particle, in quantum mechanics, is one that can never be directly detected, but whose existence does have measurable effects.

Every type of matter particle has a corresponding antiparticle. When a particle collides with its antiparticle, they both annihilate, leaving only energy.

Hawking radiation. Space is filled with pairs of particles and antiparticles; in the presence of a black hole, one particle may fall inside, leaving its unpaired mate to appear as radiation emitted by the black hole.

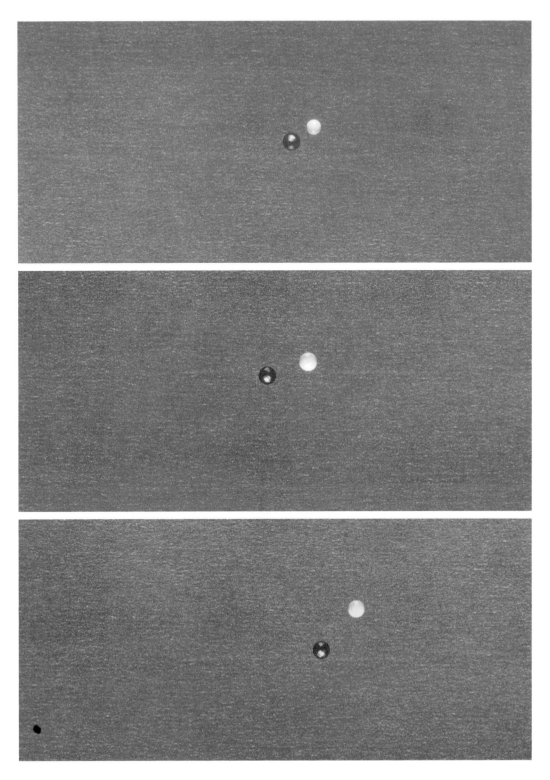

Einstein never accepted quantum mechanics, because of its element of chance and uncertainty. He said, "God does not play dice." It seems that Einstein was doubly wrong. The quantum effects of black holes suggest that not only does God play dice, he sometimes throws them where they cannot be seen.

DENNIS SCIAMA

*Y*akov Zeldovich, the famous Russian astrophysicist and cosmologist, also refused to believe it. But in a few months it was clear the arguments were correct. Apart from the breathtakingly unexpected nature of the discovery, the original calculation was rather complicated, and the effect emerged as a sort of small residual effect from a bigger phenomenon that was occurring. And as often happens in physics, once people got the idea, they cleaned up the discussion and made it more transparent. After a few months everybody agreed it was correct, and it changed our understanding of physics in a fundamental way.

Cygnus X-1 is a constellation six thousand light-years from earth that contains, many scientists believe, a black hole. Another black hole is thought to exist in the Large Magellanic Cloud, a galaxy visible only in the southern hemisphere.

The fact that this Hawking effect for radiating black holes has not been tested observationally is not, as it were, Hawking's fault. It's simply that for the sort of black holes that arise naturally, such as the possible one in Cygnus X-1, the effect would be too weak to detect.

A primordial black hole is a black hole that was created when the universe was very young.

He did have a good go at trying to find observational effects, because he suggested in one very interesting set of works that there might be primordial black holes of low mass.

You see, the temperature of a black hole is greater the less the mass of the black hole; and, of course, the hotter a black hole, the more powerful the radiation coming from it. And as it radiates and loses mass, it gets hotter and radiates faster, so the end product would be a violent explosion. In fact, Hawking calculated that that explosion would be more violent than any other known in science—except the big bang origin of the universe itself.

The whole question is: How long does this process take? Well, if you have a solar-mass black hole, and you think of it starting to radiate now, then the time it would take before it exploded so violently would be something like sixty powers of ten longer than the present age of the universe, or some ridiculously large number. But if you could have a black hole whose mass was about the same as that of a small mountain, that would go pop in about the age of the universe.

So Hawking said, perhaps primordial black holes, formed in the early universe with this kind of mass, are exploding now—we can look for those explosions both in gamma rays and perhaps in radio waves. Then serious searches for gamma rays coming from exploding black holes got exciting for a moment, because of things called gamma ray bursts whose origin is still not understood. But we now know enough about them to know that they're not Stephen's exploding black holes.

Of course, the fact that you don't see them doesn't mean his ideas are wrong. It probably means that not many primordial black holes of this low mass were formed. In any event, these effects are just too weak to observe readily in the laboratory, and for that reason they've not been tested.

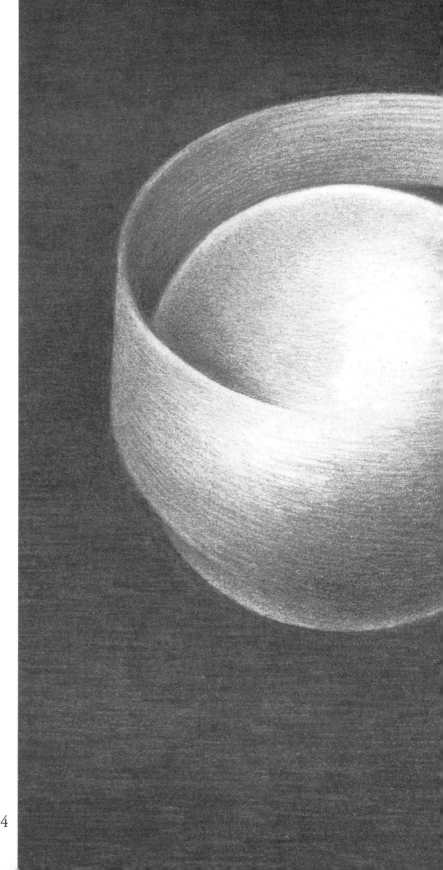

A rendition of the black hole in Cygnus X-1. Cygnus X-1 is thought to consist of a black hole and a normal star, orbiting around each other. As matter is blown off the surface of the visible star, it falls toward its unseen companion in a spiral motion, becoming very hot and emitting X rays.

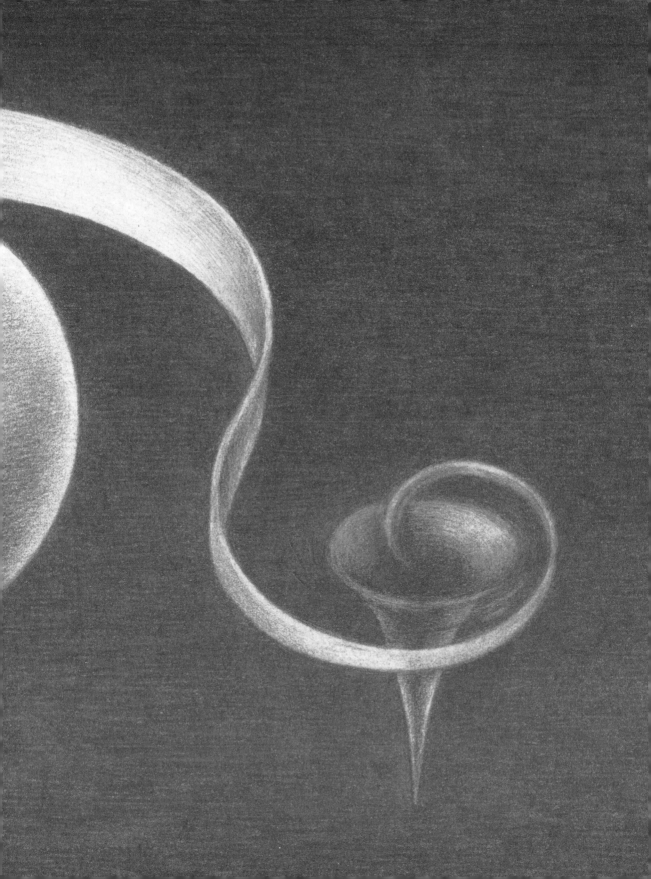

Bernard Carr

*J*ohn Wheeler once said that talking about Stephen's theory was like rolling candy on the tongue. That's true of many breakthroughs in physics. They go against the belief at the time, but once you've had them, they have that ring of truth about them.

Kip Thorne

*T*he precursor to the discovery of Hawking radiation was a meeting between Hawking and Zeldovich, at which I was present. Stephen and Jane had decided to visit the Soviet Union in 1973. Since I'm an old hand in Moscow—I've been carrying on research jointly with Soviet physicists since 1968, going and coming extensively—they invited me to go with them. And so I went.

Zeldovich was one of the key people whom Stephen wanted to meet with, and vice versa. Zeldovich was, along with Andrei Sakharov, one of the two most decorated people in the Soviet Union other than Brezhnev, whose decorations were somewhat comical. Zeldovich and Sakharov were the key designers of the Soviet hydrogen bomb. Then, in the early 1960s, they both left the nuclear weapons research effort and began to work in cosmology, black holes, and related areas.

By about 1969 Zeldovich realized that rotating black holes should emit radiation, and that this radiation should be produced by a marriage, or partial marriage, between general relativity and quantum theory. But Zeldovich believed that when this rotating black hole emitted the radiation, it would slow down the black hole's rotation and then the radiation would shut off, so basically the radiation was being produced by the rotational energy of the black hole; it was coming off from the region just outside the horizon, and emitting the radiation would stop the black hole from rotating.

Zeldovich told me in 1969 that he believed this would happen but he didn't know enough general relativity to prove it—he just knew it through intuition. I thought he was crazy, so we made a bet—he bet careful calculations would ultimately show that this was the case, and I bet that they wouldn't.

Well, when I went back in 1973 with Stephen, I paid off the bet with a bottle of White Horse scotch because it was clear by then that rotating black holes should emit such radiation.

But Stephen hadn't been aware of the idea, and when given the idea by Zeldovich, Zeldovich's explanation for it was not one he found awfully convincing—I think he wanted to think about it in his own way.

Stephen then went back to Cambridge and thought it over for a few months, and realized that even nonrotating black holes would emit radiation, and therefore a black hole could evaporate, which was far more radical than Zeldovich's conclusion that you could radiate away the rotational energy of the black hole.

Zeldovich died in 1988. He was a truly remarkable person. He was Jewish, which was part of the reason that he never traveled to the West until the Gorbachev era—the mixture of that and also his heavy involvement in the hydrogen bomb effort. He didn't understand general relativity, yet he knew intuitively that a rotating black hole would evaporate.

But his intuition did not tell him that a nonrotating black hole would evaporate. In fact, Zeldovich was the last person to give in on the issue. Two years after

Stephen showed mathematically that a nonrotating black hole must completely evaporate away, I was in Moscow and discovered that none of the Russians believed it. Why didn't they believe it? Because Zeldovich didn't believe it.

I then gave a series of lectures at one institute after another, describing Stephen's original calculation and also a different version that he had done together with Jim Hartle. These lectures forced Zeldovich to go back with Alexander Starabinsky, his brilliant young student, and rethink it in their own way.

The Sunday evening before I was to leave Moscow, I had a phone call from Zeldovich. He said, "Come to my flat. I want to talk to you."

I was madly trying to finish writing a joint paper with somebody else, but when Zeldovich calls, you go. So I flagged down a passing motorist—there are no taxis to be had, so in Moscow you flag down ordinary motorists—and he drove me to Zeldovich's flat. I went in and knocked on the door, and Zeldovich and Starabinsky were there. They threw up their hands and said, "We give in. Hawking was right. We were wrong."

STEPHEN HAWKING

Black hole radiation has shown us that gravitational collapse is not as final as we once thought. If an astronaut falls into a black hole, he will be returned to the rest of the universe in the form of radiation. Thus, in a sense, the astronaut will be recycled.

However, it would be a poor sort of immortality because any personal concept of time would come to an end as he is torn apart inside a black hole.

KIP THORNE

*B*y 1975 I was making many wagers about physics, and except for Zeldovich's, I was winning them all. So I made this bet with Stephen about whether or not there is a black hole in Cygnus X-1. I had done a moderate amount of work about that time trying to build detailed models to explain Cygnus X-1. So I made this bet that indeed it would turn out that there is a black hole in Cygnus X-1, and that modeling would not turn out to be totally useless.

The bet between Stephen Hawking and Kip Thorne on the black hole in Cygnus X-1

Whereas Stephen Hawking has such a large investment in General Relativity and Black Holes and desires an insurance policy, and whereas Kip Thorne lik to live dangerously without an insurance policy,

Therefore be it resolved that Stephen Hawking bets 1 year's subscription to "Penthouse" as agai Kip Thorne's wager of a 4-year subscription to "Private Eye", that Cygnus X 1 does not contain a black hole of mass above the Chandrasekhar limit.

Conceded
Stephen Hawking

Kip S. Thorne

June

STEPHEN HAWKING

My bet with Kip Thorne that there is no black hole in Cygnus X-1 is a form of insurance policy for me. I have done a lot of work on black holes and it would all be wasted if black holes don't exist. So, if black holes exist, Kip will get one year of *Penthouse*. If they don't, I will have the consolation of four years of *Private Eye*.

KIP THORNE

As Stephen gradually lost the use of his hands, he had to start developing geometrical arguments that he could do pictorially in his head. He developed a very powerful set of tools that nobody else really had. So in some sense, when you lose one set of tools, you may develop other tools, and the new tools are amenable to different kinds of problems than the old tools. And if you are the only master in the world of these new tools, that means there are certain kinds of problems you can solve and nobody else can.

ISOBEL HAWKING

He says himself that he wouldn't have got where he is if he hadn't been ill. And I think that's quite possible. As Samuel Johnson said, the knowledge you're to be hanged in the morning concentrates the mind wonderfully. And he has concentrated on this in a way I don't think he would have done otherwise, because he always took a great interest in a lot of things in life and I don't know that he'd ever have applied himself in the same way if he'd been able to get around. So in a way . . . no, I can't say anyone's lucky to have

110

an illness like that, but it's less bad luck for him than it would be for some people, because he can so much live in his head.

STEPHEN HAWKING

*U*p to 1974 I was able to feed myself and get in and out of bed. Jane managed to help me and bring up two children without outside help. However, things were getting more difficult, so thereafter we took to having one of my research students living with us.

DON PAGE

I would usually get up around seven-fifteen or seven-thirty, take a shower, read in my Bible, and pray. Then I would go down at eight-fifteen to get Stephen up. At breakfast I would often tell him what I'd been reading in the Bible, hoping that maybe this would eventually have some influence.

I remember telling him one story about how Jesus had seen the deranged man, and how this man had these demons, and the demons asked that they be sent into a herd of swine. The swine then plunged over the edge of the cliff and into the sea. Stephen piped up and said, "Well, the Society for the Prevention of Cruelty to Animals wouldn't like that story, would they!"

Another time Jesus was talking about the last days, and

Don Page studied under Kip Thorne at the California Institute of Technology, where he received a Ph.D. in physics and astronomy in 1976, and where he met Stephen Hawking, who was spending a year there as a visiting scholar. Don Page is an evangelical Christian; he believes that an understanding of this universe may reveal some aspects of God, but that there is more to God than this universe alone. He lived with the Hawkings from 1976 to 1979, while doing his postdoctoral work at Cambridge. He is now a Professor of Physics at the University of Alberta in Edmonton.

he said at those times there will be two working in a field and one will be taken and one will be left, and there'll be two in bed and one will be taken and one will be left. At breakfast Stephen said, "And two at breakfast, and one will be taken and one will be left." So he understood the point of the story; he took it in a good-natured way.

Then we would work. Usually we'd see if there were any scientific papers people had sent out—the papers you asked other people to read before it was time for them to get published in the scientific literature. I would sort of hold the things in front of Stephen because he couldn't turn the pages, and that would give me a chance to read them, too.

Don Page and Stephen Hawking at work

I discovered that in spite of Hawking's great brains, he does read quite slowly. I could read about twice as fast as he. But of course the point is that he would have to read to remember it because it would be very difficult for him to go back and access the thing. Whereas I could just skim the paper rather quickly and see if there was something interesting in this, and if I wanted to work on it, I could go back and pick it up and look at it again.

KIP THORNE

Once I visited Stephen and Jane at their home in Cambridge, and after supper, when it was time for Stephen to go to bed, he made his way up alone. I have forgotten whether it was one or two flights of stairs. This was a period when Stephen could no longer walk—the way he got up the stairs was by grabbing hold of the pillars that supported the banister and pulling himself up with the strength of his own arms, dragging himself up from the ground floor on up to the second story in a long, arduous effort.

Jane explained that this was an important part of his physical therapy, to maintain his coordination and strength as long as possible.

It was heartrending to watch what appeared to be agony until I understood. As it is with so many things about Stephen, once you get to know him you take for granted things you should take for granted—it's just the way things are. Yet at first, when you're a stranger to it, you're really taken aback and you see him through very different emotional eyes than after you've come to know him well. It's simply part of life, pulling himself up the stairs like that.

*On a Cambridge street.
"He goes dashing about all
over the place."*

ISOBEL HAWKING

One of the worst things for me would be having people
there all the time. I couldn't bear that. Yet he finds things
funny, and he enjoys life, and he goes dashing about all over
the place. I think this is tremendous. It is the sort of courage
that I haven't got and his father hadn't got. And we cannot
but admire it, but wonder how on earth he got it.

DON PAGE

One year the Hawkings took me along when they went
to a cottage in Wales, near the river Wye. This cottage was
up a hill, and there was a bit of paved sidewalk that went to

114

it. Of course, I wanted to do it in the least number of trips so I put extra batteries under his chair. Stephen, however, didn't realize that I'd put them under there, so he didn't know that his wheelchair was as heavily laden as possible.

So he started up this slope and got quite a bit ahead of me, about ten meters, and then he turned the corner to go into the house—but this was on the slope. I looked up and I noticed Stephen's wheelchair was slowly tipping backward. I tried to run up, but I wasn't able to move nearly rapidly enough to stop him from toppling over backward into the bushes.

It was a bit of a shocking sight to see this master of gravity overcome by the weak gravitational force of earth.

KIP THORNE

*O*n a couple of occasions in the early to mid-1970s, Stephen and I had discussions of what his future was. He seemed to have a very clear picture at that time. He said he expected that he would ultimately die as a result of catching pneumonia. He didn't know when this would happen, but that it would happen—but also that he expected there would not be any degeneration of his mental abilities before it took place—he was quite confident of that. And he seemed quite matter-of-fact about the idea that this was his ultimate fate.

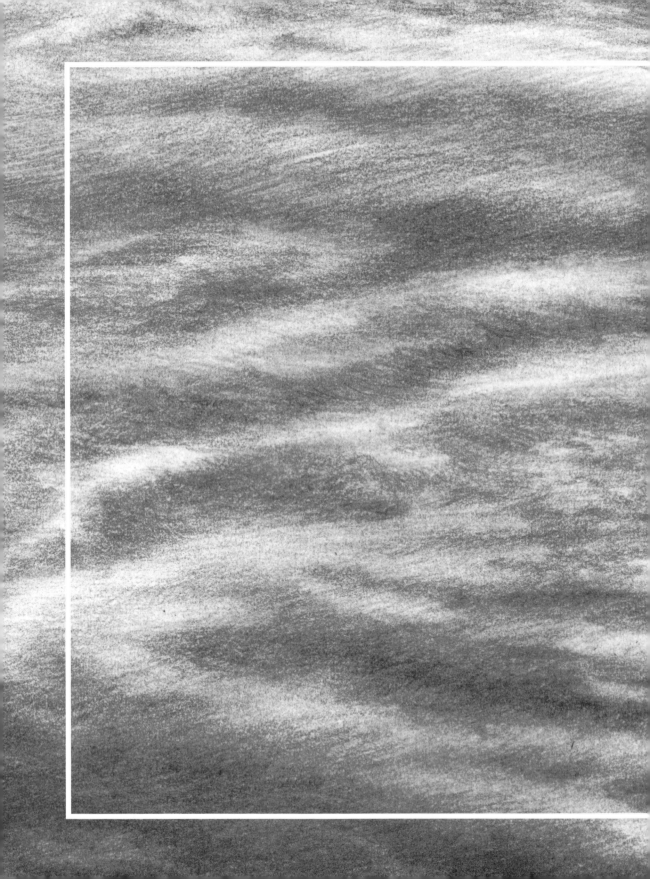

FOUR

In 1975 the Vatican awarded its Pius XII medal "to a young scientist for distinguished work" to Stephen Hawking, and he flew with Bernard Carr to Rome to receive the award in the Vatican from Pope Paul VI.

BERNARD CARR

*I*t was a very moving occasion. Normally, the person getting the medal would go up to the Pope to receive it, but of course Stephen was disabled, and so the Pope came down to Stephen.

The point is historically that there has been something of a conflict between the Church—the Catholic Church in particular—and the history of cosmology, going back to the days of Galileo. Stephen has a great affinity with Galileo. I remember, when we went to the Vatican, he was very keen to go into the archives and see the document which was supposed to be Galileo's recantation, when he was put under pressure by the Church to recant on his theory that the earth went around the sun.

The controversy between science and the Church still rages. I think it gave us some pleasure that the Church finally announced that they'd made a mistake with Galileo, and that in fact Galileo was right. But whether or not the Pope would have approved of what Stephen had discovered, if he actually had understood it, I am not quite sure.

KIP THORNE

*T*he thing that stands out most clearly in my mind is that there's been a marked change in Stephen's style of research since the early 1970s, and it's a change characterized by words he said to me around 1980: "I would rather be right than rigorous." Rigor is something that mathematicians seek—they want to have a firm, clear mathematical proof that they're correct. Stephen sought that kind of rigor in the 1960s and early 1970s in his research; he tried to make sure that everything was completely firm.

In more recent years he's become much more speculative, seeking the truth; wanting to be, let's say, 95 percent sure he's right, and then to move on quickly. He has abandoned the quest for certainty which he seemed to be seeking in the early 1970s in favor of high probability and rapid movement toward the ultimate goal of understanding the nature of the universe.

STEPHEN HAWKING

My interest in the origin and fate of the universe was reawakened when I attended a conference on cosmology in the Vatican in 1981. Afterwards, we were granted an audience with the Pope, who was just recovering from an attempt on his life. He told us that it was all right to study the evolution of the universe after the big bang, but we should not inquire into the big bang itself because that was the moment of creation and therefore the work of God. I was glad then that he did not know that the subject of the talk I had just given at the conference was the possibility that space-time was finite but had no boundary, which means that it had no beginning, no moment of creation.

It was not immediately obvious that my paper had implications about the origin of the universe because it was rather technical, and had the forbidding title, "The Boundary Conditions of the Universe." In it I suggested that space

and time were finite in extent but were closed up on themselves without boundaries or edges, just as the surface of the earth is finite in area but doesn't have boundaries or edges. In all my travels, I have not managed to fall off the edge of the world.

At the time of the Vatican conference, I didn't know how to use this idea to make predictions as to how the universe would behave. But during 1982 and 1983, I worked with my friend and colleague Jim Hartle, of the University of California at Santa Barbara. We showed how to use this no boundary idea to calculate the state of the universe in a quantum theory of gravity.

If the no boundary proposal is correct, there would be no singularities, and the laws of science would hold everywhere, including at the beginning of the universe. The way the universe began would be determined by the laws of science. I would have succeeded in my ambition to discover how the universe began. But I still don't know why it began.

Jim Hartle, Stephen Hawking's collaborator on the Hartle-Hawking proposal, is currently Professor of Physics at the University of California at Santa Barbara, where his work focuses on relativity and gravitation.

JIM HARTLE

In quantum mechanics, we describe a system by giving its wave function. That enables us to compute probabilities of what we might see. In the case of the universe, for example, we might be interested in its size and shape and the three-dimensional geometry of space that it might exhibit. The wave function allows us to calculate probabilities for different alternative answers. So we've been investigating the possible modes of description. But we don't live in such a description, we live in one quantum state, and therefore the interesting question is: What are the principles which would single out one state from among the many possible states that the universe might have, and therefore give us a mechanism for making predictions or correlating different features of the universe which we see today?

DENNIS SCIAMA

Stephen's "quantum cosmology" or, more popularly, "the wave function of the universe," is a new physics proposal. It says that the way to calculate the behavior of the universe using quantum mechanics is to use certain highly technical procedures which represent a new proposal for the way physics works; therefore they're controversial and not yet definitely established.

But what he tries to show is that if the wave function of the universe is calculated by a procedure which he's suggested—a very clever and brilliant suggestion but not one that everyone agrees is likely to be successful—then you can work out what the universe would be like. He and his students and others elsewhere in the world test this new theory by seeing whether the real universe has the same properties as the theoretical implications of his proposal. His view is that this is leading to a successful set of ideas. But it all remains controversial at the moment.

In classical physics, one can describe the state of a system by giving the positions and speeds of all the particles at one time. In quantum physics, on the other hand, particles do not have precisely defined positions and speeds. Instead, the most complete description one can give is what is called the wave function. This can be thought of as giving the probabilities for finding the particles in different positions. One does not also have to specify the speeds of the particles. These are determined up to a prescribed amount of uncertainty by the wave function.

In quantum theory, the state of the universe can be described by what is called the wave function of the universe. This gives the probability for space at the present time to be curved or warped in different ways. The probability is greatest for space to be nearly flat, but the no boundary proposal of Hartle and Hawking predicts that there are significant probabilities for small ripples in the geometry of space. These would correspond to gravitational waves, and very sensitive measurements are now being carried out to try to detect them.

KIP THORNE

*T*here are many different approaches to quantum gravity, to the marriage between general relativity and quantum theory. The one I find most appealing is the approach that Hartle and Hawking have taken with their descriptions of how quantum mechanics and gravity are married. To me it smells right. When you're working at the ultimate frontiers of science, of physics, you have to rely in great measure on how something smells, on how something feels. You have to make decisions about where you go in your own research, or what kind of guidance you give to your students about what are the fruitful directions. That kind of guidance has to come from intuition, from gut feeling. As Soviet colleagues say, it comes out of the liver. Well, I feel in my liver that they are right.

ANTONY HEWISH

I think the most fantastic thing about the last twenty years is that you can now argue about what went on before a millionth of a second after the beginning of time. That the universe has a singular origin is a staggering concept, but the fact that you can talk sensibly about the physics of that early universe and debate the first millionth of a second strikes me as being such a revolution, particularly when I look at it over my own life span, in which at first we didn't even know whether the universe was evolving or steady state.

And it does raise these dreadful philosophical questions such as, did time have a beginning, and what does that really mean? It's hard for physicists really to get to grips with that sort of problem.

John Wheeler

*M*any times, by many people, in many contexts, the question has been raised about the beginning of the universe: How did it begin?

Einstein himself didn't trust the predictions of his own theory when he first explored them and found they said that the universe could not stay fixed in size for all time. Why was he so distrustful? Because his greatest hero, Spinoza, had in effect argued long before then against the biblical idea of an original creation; because Spinoza said in effect, "Where would the clock exist before anything was there to tell it when to begin?" It seemed like a contradiction in terms.

But of course, when in the end it turned out that the universe is expanding, Einstein told his and my great friend George Gamow, "That was the biggest blunder of my career."

And ever after, he and I certainly look at that prediction that the universe expands—something so totally different from anything one had ever conceived before—as the most powerful evidence man has ever been granted of our ability to look into space, into the way of working of the universe, and our greatest token that we can someday solve the mystery. But I think, in the light of quantum theory, we have a deeper look at all such issues, of how the universe began, because there we come to realize that the very idea of time is an idealization. The word "time" was not handed down from heaven as a gift from on high; the idea of time is a word invented by man, and if it has puzzlements connected with it, whose fault is it? It's our fault for having invented and used the word.

In the men's room of a café in Austin, Texas, I once saw a bit of graffiti: "Time is Nature's way to keep everything from happening all at once"—which is a shorthand way to speak of how we usually understand time. But the deeper we get into time from the quantum point of view, the more insights we get, the more sophistication time itself develops, and the heavier become the puzzles that we have to solve to make headway into a thorough understanding of the whole show.

So anybody who asks questions about how the universe began ought to have asked of him in return, "Where did you get your idea of time?"

⸻

In quantum mechanics, the position of an electron is uncertain. Under normal circumstances, this uncertainty is negligible, but it becomes important on atomic distance scales, so much so that a question such as "Where is the electron?" has no meaningful answer. Likewise, in quantum gravity, there is also a certain uncertainty in the geometry of space-time. Under normal circumstances, this uncertainty is also negligible, but it becomes important at very small distance scales and very small time intervals, so much so that a question such as "What time is it?" has no meaningful answer. In particular, this means that very close to the big bang, the concept of time becomes meaningless.

Stephen Hawking suggested that progress might be made by using imaginary time. Imaginary time is related to real time in the same way that imaginary numbers are related to real numbers. Imaginary-time physics is an Alice-in-Wonderland version of physics; for example, particles can move faster than light and even backward in (imaginary) time. Of course, when the description is turned into an approximation to quantum mechanics, everything is made right, and nobody can go faster than light or backward in (real) time.

Jim Hartle and Stephen Hawking proposed a method for implementing the imaginary-time trick in quantum cosmology, and used it to analyze a drastically simplified model of the universe. Although much simpler than the real universe, the model did possess a big bang singularity when it was treated with classical general relativity. But the quantum-mechanical version had no singularity at all. Hartle and Hawking had succeeded in replacing the singularity with what could be called a prehistory of time.

It is still an open question as to whether the Hartle-Hawking idea can be successfully applied to a real universe.

Christopher Isham is Professor of Theoretical Physics at Imperial College, London; he is working on the relationship between general relativity and quantum physics.

CHRISTOPHER ISHAM

The concept of creation out of nothing is, of course, a fascinating one; it's something which interests people enormously. The Hartle-Hawking picture of creation from nothing is actually a very graphic way to describe the mathematics in nontechnical terms. You see, what normally happens in physics, where you have conventional ideas of causality and determinism, is that if you are given the state of some particular time, you can work out uniquely the state of some later time. That's what one means by causality. From that point of view, you could never have a creation, as it were, out of nothing; in fact, you can never have creation at all. All you can really have is change: causal change, but change in what is already there. This change may appear to be involved in creation—as, for example, in the particle accelerators at CERN you see elementary particles banging together and floods of new ones appearing; that looks like creation, but all you are really doing there is converting energy from one form into another, and the whole system is purely causal, purely deterministic—everything just flows in the conventional way, and certainly you aren't creating from nothing.

CERN is the European Organization for Nuclear Research, near Geneva, Switzerland.

In general relativity, due to the way that time and space enter into the formulas, the possibility does arise of actually talking about the creation of time itself. But the trouble is that when space and time appear to "come into being" in the classical theory, that actual point is itself a singular point in the mathematics—the mathematics breaks down, so you cannot use it to give you a creation theory. All you can say in conventional cosmologies is that there exist many, many different possible universes, all of which are consistent with Einstein's equations. The fact that we happen to live in this one rather than that one is pure coincidence. You can give no reason for it—even in principle you can't. All you can say are conditional statements: given that the universe was in this state at this time, then it will be in that state at a later time. It's a conditional type of evolution.

However, when you come to imaginary time, you have this rather peculiar possibility of having a "now," as it were, without necessarily having a chain of past moments. If we start where we are at the moment and run backwards in time, for a long while things work perfectly normally, even when you have imaginary time. As long as you use this phenomenological time, it looks as if you're just sort of tracking backwards in conventional time.

But as you begin to get further and further back towards what looks as if it would be the origin point in the conventional real-time picture, you find that the nature of time changes—the complex or imaginary component becomes more and more prominent until, in the end, what ought to have been the singular point in the classical theory just gets smoothed away, and you have this beautiful picture, these sort of bowls of the creation of the universe, where there's no initial point, just a sort of smooth shape.

What Hartle and Hawking discovered was, if you supposed that the past history of the universe in imaginary time is all possible shapes like this which happen to fit into where we are at the moment, and you interpret this in a more or less conventional quantum-mechanical way, what you end up with, at least in principle, is a unique wave function for the whole universe.

So you have this nice picture of no past—there's nothing out of which the universe was created at all. All you can really say is that the universe *is*, because it's a self-consistent mathematical structure. There's no past because, unlike the creation-at-a-point scenario, there's nothing for it to be created in.

So to say the universe is created from nothing is actually a little bit of a misnomer; it's a misleading use of the word "nothing." It's not just that there was empty space in which the universe appeared, which you might call nothing: there really was nothing at all, because there wasn't even a creation event.

The use of the past tense in a verb becomes inappropriate in these theories. Tenses were set up when people believed in real time, of course, and unfortunately we don't yet have

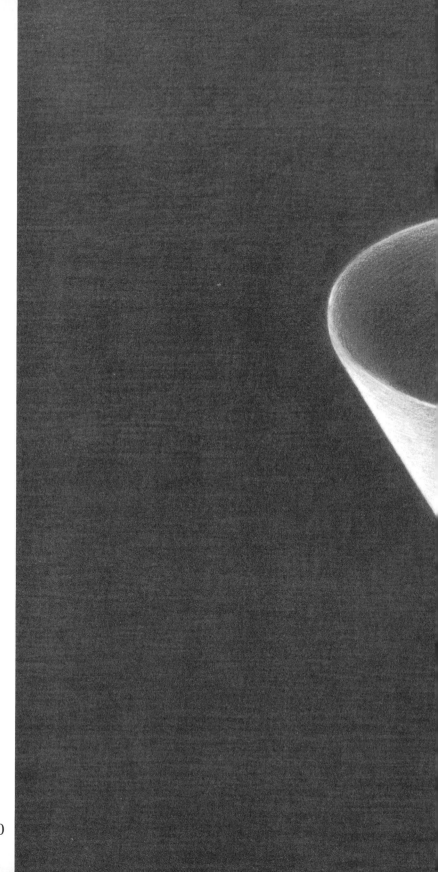

The open universe: "As you get further back, the nature of time changes . . . until you have these bowls of the creation of the universe, where there's no initial point, just a sort of smooth shape."

130

a linguistic form to describe tenses in imaginary time. But certainly, in that sense, the use of the term "creation from nothing" is misleading. It would be appropriate for this picture of the universe suddenly popping out in a pre-existing time, but it's not really a very good description of the Hartle-Hawking state.

STEPHEN HAWKING

*I*n order to predict how the universe started off, one needs laws that hold at the beginning of time. In real time, there are only two possibilities: either time continues back into the past forever, or time has a beginning at a singularity. One can think of real time as a line going from the big bang to the big crunch. But one can also consider another direction of time at right angles to real time. This is called the imaginary direction of time. In the imaginary direction of time, there need not be any singularities that form a beginning or an end to the universe.

In imaginary time, there would be no singularity at which the laws of science broke down, and no edge of the universe at which one would have to appeal to God. The universe would be neither created nor destroyed. It would just be.

Maybe imaginary time is really the real time and that which we call real time is just a figment of our imaginations. In real time, the universe has a beginning and an end. But in imaginary time, there are no singularities or boundaries. So maybe what we call imaginary time is really more basic and what we call real time is just an idea that we invent to help us describe what we think the universe is like.

KIP THORNE

*T*here are two basic theories of how the universe will end. There's the idea of the open universe which will go on without any sudden end—things will merely slow down and come to a heat death according to the second law of thermodynamics. Alternatively, there's the idea of the closed universe which will stop expanding and fall back on itself—this is sometimes called the big crunch, like the big bang, only the other way around.

JIM HARTLE

*I*t's important to understand that the word "imaginary" in imaginary time doesn't refer to imagination: it refers to a very ancient idea in mathematics, namely, imaginary numbers, such as the square root of -1. For a given observer, space and time are of course distinct: we measure space with rulers and we measure time with clocks. Early in this century, Einstein and Hermann Minkowski showed that the notions of space and time appropriate to different observers were just different aspects of one unified idea, the notion of space-time. Space-time is a four-dimensional geometry which has some space-like directions and some time-like directions. So, in a certain sense, the notions of space and time are still distinct there.

Despite the great power of that idea, it's still possible to go further in unifying these notions. If you measure time directions using imaginary numbers, then you obtain complete symmetry between space and time, which is, mathematically, a very beautiful and natural idea. This mathematical simplicity is exploited in the no boundary

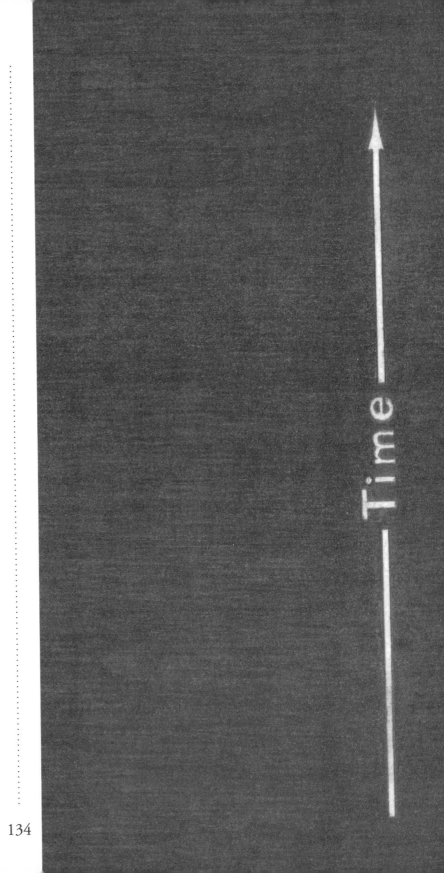

"One can think of real time as a line going from the big bang to the big crunch."

134

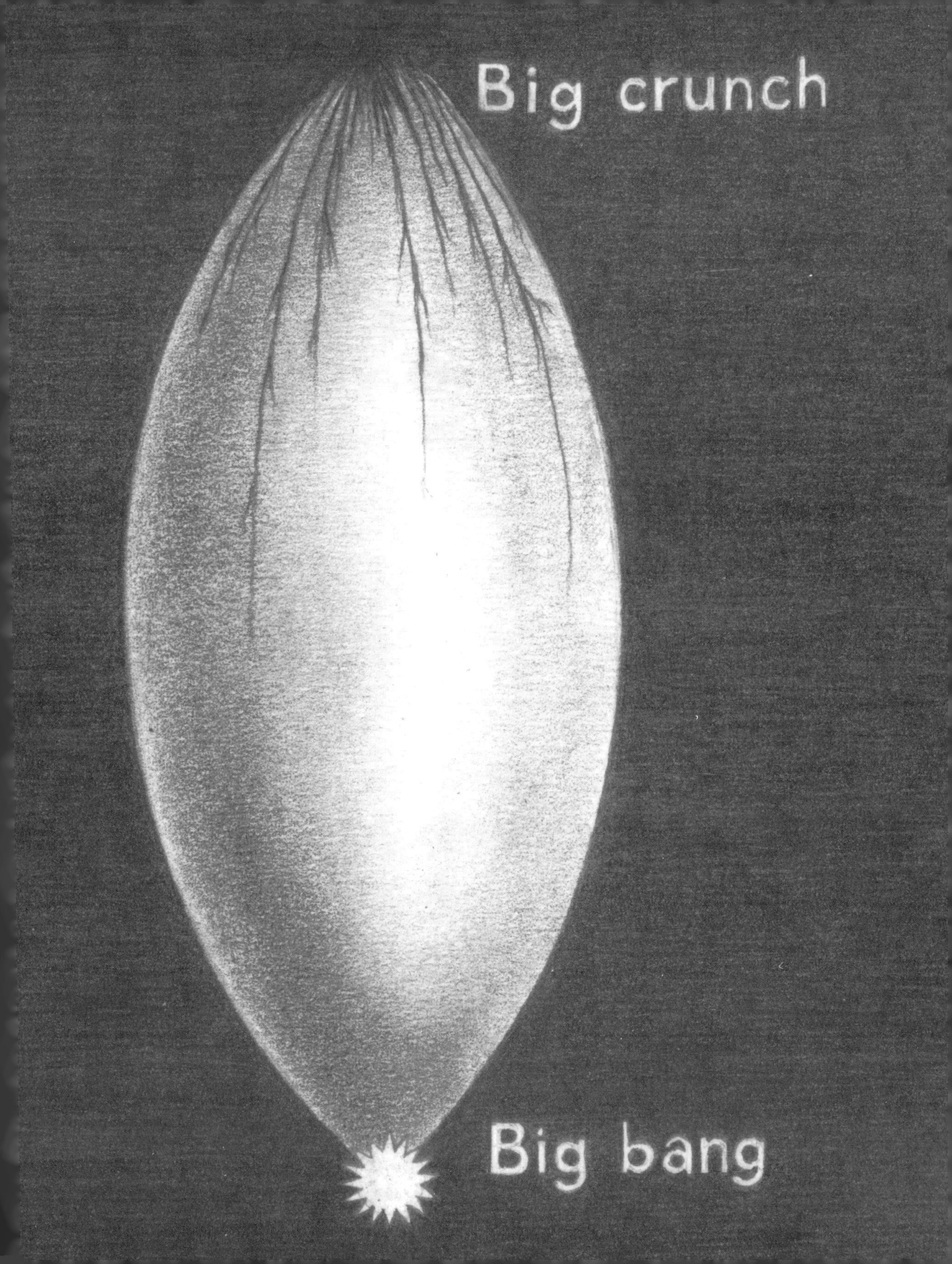
Big crunch
Big bang

proposal to give a theory of the simplest of all possible initial conditions of the universe.

But one shouldn't think of imaginary time as something to which we have direct access in our experience: it's a mathematical idea which expresses the beauty of the equations of physics, and in this case a particular proposal for the initial condition of the universe.

DON PAGE

*H*awking's singularity theorems indicated that Einstein's theory of general relativity, combined with certain observations, implied the universe had a singularity at the beginning. If you extrapolate back in time, you get to some point beyond which you can't go. We usually interpret that to say that time began there.

This upset a lot of people who had assumed the universe was infinitely old. Hawking's idea indicated the universe had a beginning, and some felt that fit with the Genesis accounts of the universe having been created in time; though of course other theologians say that God's creation doesn't necessarily imply that it's within our time.

God could have created an infinite universe, but Hawking's idea implied a beginning to time. Now, of course, we know that Einstein's theory does not work very close to that beginning, that the theory itself breaks down there. That does raise the possibility that maybe the universe was infinitely old, or that maybe something else applies.

Now, a number of people, including myself and, I think, Hawking, feel the concept of time itself breaks down near the beginning, and therefore it may make no sense to talk about what was before that: Was there an infinite time before? Was there a finite time? Did the universe have an absolute beginning in time? Some of those questions just don't make much sense because the concept of time itself doesn't make much sense at these really early times. All we can say with fair confidence is that time as we know it had a beginning, but that there was a point beyond that where our standard notions of time break down.

In the Hartle-Hawking no boundary proposal the universe essentially starts in a way in which time has a funny behavior: time is in a technical sense imaginary, so that instead of an edge to time, it's as if you have the surface of the earth and you start from, say, the North Pole and come outward, along the lines of longitude. These lines of longitude indeed start at the North Pole, where it's perfectly regular.

This is in a way Hawking's picture of the universe: that this imaginary time had no beginning, it had no edge. It didn't necessarily go on forever. It was finite, just in the same way that there's only a finite amount of area in the earth, which doesn't go on forever as you go north—in some sense it comes to an end, since there is a farthest north you can go. But in another sense there's no real end there.

So Hawking is saying that the universe had no boundary at the beginning, and therefore the universe is a self-contained whole. And he argues that God didn't really have to start off the universe: the universe could just exist on its own, without God's creating it.

STEPHEN HAWKING

*M*ost people have come to believe that God allows the universe to evolve according to a set of laws and does not intervene in the universe to break these laws. However, it

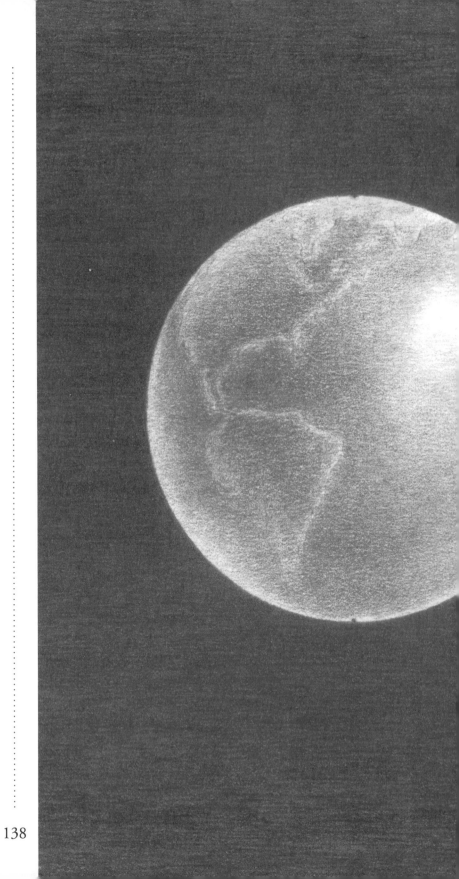

The no boundary proposal: the history of the universe may resemble the surface of the earth; it may start at a single point at the equivalent of the North Pole and continue to the South Pole, but there is no beginning, nor an end.

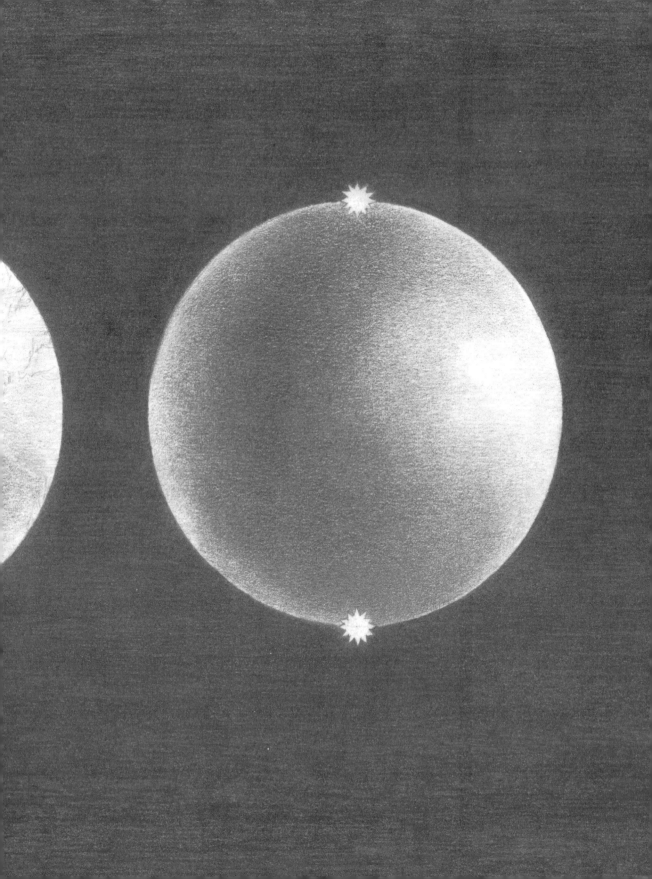

would still be up to God to wind up the clockwork and choose how to start it off. So long as the universe had a beginning, we could suppose it had a Creator. But if the universe is really completely self-contained, what place then for a Creator?

DON PAGE

What place then for a Creator? The question as to whether God created the universe is not directly related to whether the universe has an edge, even though many people think it does. It's actually somewhat irrelevant.

For example, I've drawn a couple of lines on a piece of paper. This straight line here has two ends: you could say this end is a beginning and this one is an end, if I imagine time going that way. If time goes the other way, then it's the reverse: this end is the beginning and that's the end. You might think of this as one model for the universe, a universe that has a beginning and has an end.

Then there's another universe shown in this circle, where, as time goes across here, in some sense there's an earliest time; but if you follow the line in a circle, the line has no end, it just goes around.

But I myself drew both of those lines, so in a sense I created them both. Yet the question of whether I created them makes no difference as to whether they have a beginning or an end.

I think it's similar for the universe. Hawking's old model gave the universe a beginning and perhaps an end. The new model is more like this circle in which there's not really a beginning or an end. There is in some sense a farthest to the left; so you could say there's something like an earliest time and there's something like a latest time. But in a more technical sense there's no beginning and no end. And yet these both could have been created by God. It's through faith that we can ask the question of whether it was created by God. That is a question that science can neither affirm nor refute.

In his book, I think Hawking is careful not to come out and say directly there is no God. He just says: What place then for a Creator?—although I think it's pretty obvious what conclusion he'd like you to draw from it.

JOHN TAYLOR

I feel God is very important to many people because of the need to explain both the enormous complexity and the possible lack of purpose in the universe. I would never wish to remove that great aid it gives people through this life.

Personally, I don't see God as being related to anything in more detail than a possible first cause. But being a pragmatic materialist, I would say that God is to be described in terms of the laws that would determine that first cause. How you do that, I don't know. We're not at the stage where we can say we have a glimmer of understanding.

I feel that we need to move on and see more closely the nature of God—because if you ask me, "What do you feel is God?" I would say, "It is the nature of the universe." And I feel that perhaps we will always move on: there will always be another theory behind the present theory, there will always be another step to be taken. In a sense, that may be a way of avoiding having to reach a first cause.

So if a scientist were to say, "We've come to the end," I would reply, "Think very carefully. Maybe the nature of the universe is such that it is infinitely complex." Possibly there could be a theory of infinitely complex systems in which there are infinite sequences of theories going ever back to ever shorter distances. Maybe we can get a supertheory; I don't know. In fact, you might say that if God is the ultimate nature of the universe, as expressed in theory, there must somehow be a theory of that to explain how it can be controlled.

At that point, there are such deep logical questions, as well as the possibility that we are not far enough along the road towards the sequence of theories to begin to see structure in it, that perhaps the truth is we should not try and run before we can walk.

The purpose in the universe, it seems to me, is to progress in the way required to progress, and only in that way; and to bring into mind a form of human purpose—that purpose being to achieve some goal that satisfies the universe's dynamical laws. I see nothing else in the way of purpose.

ROGER PENROSE

I'm not sure that the word "purpose" one might use in connection with the universe or the laws of physics is quite the same as the way we use the word in a personal sense: when we intend to do something. But there is a certain sense in which I would say that the universe has a purpose. It's not just there by chance.

Some people take the view that the universe is simply there and it runs along — it's a bit as though it just sort of computes, and we happen by accident to find ourselves in this thing. I don't think that's a very fruitful or helpful way of looking at the universe. I think that there is something much deeper about it, about its existence, which we have very little inkling of at the moment.

JOHN WHEELER

*H*ere I was, coming into the room, and all my friends had big grins on their faces. I knew they had some trick in mind. Still, I started in:

"Is it animal?"

"No."

"Is it vegetable?" I asked the next one.

"No."

"Is it mineral?" I asked the third one.

"Yes."

And then I asked the next one, "Is it green?"

"No."

"Is it white?"

"Yes."

I kept on but I noticed my friends were taking longer and longer in answering. I just couldn't understand why, because they had the word, so why couldn't they tell me right away yes or no?

I knew I only had twenty questions and pretty soon I had to make up my mind on some word. So finally I asked a friend, "Is it cloud?"

He thought and thought and thought until he said, "Yes." Then they all burst into laughter. They explained that when I went out of the room they had not agreed on a word; they'd agreed to agree on no word. Everyone could answer my question as he wished, with one proviso: if I challenged and he couldn't answer, he lost and I won. So it was just as hard for every one of them as it was for me. That word in the room didn't exist until I came in; it was only through my choice of questions that it came into being, yet it existed not through my questions alone, but through their responses, too.

And so it is with the electron: we used to think the

electron existed in the atom with a position and a velocity. Now we learn that it does not have a position or a velocity in the atom. Not till I install the measuring equipment and measure the one or the other do I get an answer. So inevitably the world has this participatory character—we are not simply observers, we are also participators in bringing about that which we have the right to say has happened.

This participatory character of the universe is the most challenging feature of the quantum, and the one that's the most exciting to try to explain, because it's the most contrary to the old idea that, as Einstein thought, the universe exists out there independent of us. Einstein was uncomfortable with the idea of quantum theory that somehow we are involved in bringing about that which we say has happened. Uncomfortable with it because, after a bitter fight within himself as a young man, he had come to the view that the world is something that exists out there independent of us. Here, in the quantum, was something totally contrary.

I had worked with the other great man in the quantum debate, Niels Bohr, in Copenhagen. And I know no greater debate in the last hundreds of years than the debate between Bohr and Einstein, no greater debate between two greater men, or one that extended over a longer period of time—twenty-eight years—at a higher level of colleagueship. To put it in brief: Does the world exist out there independent of us, as Einstein thought; or, as Bohr thought, is there some sense in which we, through our choice of observing equipment, have something to do with what comes about? That debate most people in the world of physics—myself included—feel was settled conclusively in favor of Bohr.

Today, for many people in the world of physics, quantum theory is a magic sausage grinder: you put in your theory of the solid state; you put in your treatment of the atom; you put in your treatment of the laser; you put whatever the physical problem is into this overarching principle of twentieth-century physics, this sausage grinder, you turn the crank, and out comes the answer.

In fact, quantum theory is not something that lends itself to simple analysis. Quantum theory is something that is inescapable. It shows us that what we say happens, or what we have the right to say happens, is inescapably dependent on what choice of measurement we choose to make. This choice is irretrievable, there's no opportunity to reverse it, so that we have here a revolutionary side of the story of existence.

In the very last interview he gave in his life, Niels Bohr—happy, interested, charming, concerned, destined to die the next day but not knowing it—remarked that certain philosophers did not realize that in quantum theory something really new about the world had been discovered, something, as he put it, "of very great importance." In other places he had written about how our view of reality is different than we had thought, and how we're destined for a revolutionary change. Einstein said in the most discussed paper that he ever wrote that quantum theory, as he understood it, was contradictory to every idea of reality that was reasonable. Bohr replied to this, "Your idea of reality is too limited."

JIM HARTLE

*A*nyone over the age of twelve knows that there is no such thing as certainty in this world, right? And therefore physics has to deal in probabilities.

It was the vision of classical physics that we dealt in

probabilities because of ignorance; that is, that we simply didn't have a sufficiently refined and accurate description of the world, but that when we found that description, then we would have certainty. Now, for some sixty years, we've known that this vision is false. Probabilities are fundamental, uncertainty is inevitable, and therefore a quantum-mechanical theory of anything—the universe in particular—doesn't predict one particular time history of the universe. Rather, it predicts probabilities for various possibilities that might have happened.

Some things, of course, are much more probable than others, and the rules of quantum mechanics tell us what these things are. And so, in the history of the universe, what we're talking about is all the events right back to the beginning. There's not one set of events that's legislated by the laws of physics; all events are possible, some more probable than others. It's the task of quantum cosmology to identify those things which are predicted by the theory with very high probability.

The different alternatives for the universe, the different alternative histories, are sometimes—I think not incorrectly—referred to as being equally real. A quantum-mechanical theory that deals with probabilities necessarily deals just with sets of alternatives, and therefore one can say that they're all equally real. But it would be more precise to say that we're dealing with a whole set of possibilities, any one of which could have occurred, and some of which are much more likely than others.

STEPHEN HAWKING

*E*instein once asked the question: How much choice did God have in constructing the universe? If the no boundary proposal is correct, he had no freedom at all to choose initial conditions. He would only have had the freedom to choose the laws the universe obeyed.

This, however, may not have been all that much of a choice. There may well be only one unified theory that allows for the existence of structures as complicated as human beings who can investigate the laws of the universe and ask about the nature of God.

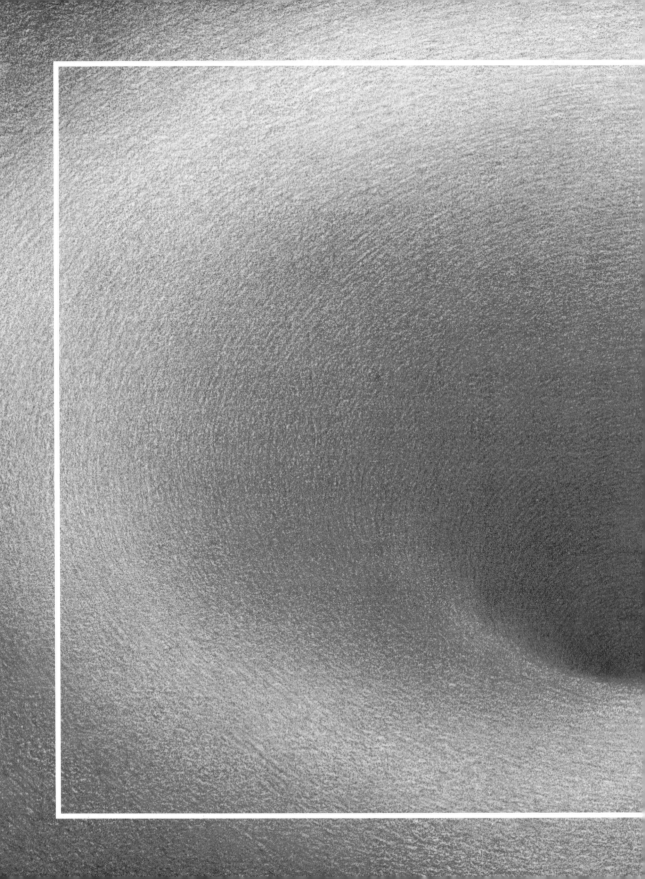

FIVE

is supposed to sign. After I had been Lucasian Professor for more than a year, they realized I had never signed. So they brought the book to my office and I signed with some difficulty. That was the last time I signed my name.

BERNARD CARR

$\mathcal{I}$ had been in awe of Stephen as my supervisor—one is always a bit in awe of one's supervisors. And when the supervisor is Stephen, one is obviously in more awe.

On the other hand, I got to know Stephen quite well, because he became a friend when I lived with the family for a year. In those days, Stephen could still talk, but in a way that was difficult for people who didn't know him well to understand. He was still giving seminars, and by making a lot of effort he could make himself understood by people; but what would tend to happen was that when we went to meetings, Stephen's students or his family would often, in a sense, have to interpret.

It became more difficult as the years went by. Once Stephen left a party and someone was helping him down the stairs. Stephen was trying to say something to this person, but the man couldn't hear it, so Stephen kept on repeating himself. This person became very worried and thought, "Goodness, this might be a very serious situation. Maybe Stephen is very ill." So he put him down and rushed upstairs and said, "Quick, come and help me. Stephen needs help!"

So everyone went down the stairs and somebody was able to interpret what Stephen was saying—he was just telling the punch line of a joke.

In 1982, when Stephen Hawking was faced with the fees of his daughter Lucy's new school, he decided to write a book about the universe directed toward readers without a scientific background. By 1984 he had completed a first draft of A Brief History of Time *and was working on revisions. Then, while on a visit to Geneva, Switzerland, he developed pneumonia and had to undergo a lifesaving tracheostomy that removed his ability to speak.*

Brian Whitt did his Ph.D. on quantum gravity from 1982 to 1985 under Stephen Hawking and then spent three more years working with Hawking as a Research Fellow. Today he runs his own computer business in Cambridge.

Brian Whitt

*H*e fell ill in Switzerland. When he came back, he was on a ventilator, so he had a tube down his throat. From that point on, he couldn't speak.

I can't remember how long he was in intensive care in Cambridge. But during that period, maybe a couple of months, I spent probably one in two or three nights a week at the hospital because he couldn't communicate with the nurses. It's not just like being seriously ill—the nurses simply couldn't understand what Stephen wanted. If he was uncomfortable, they couldn't tell why. And so a number of us were there effectively twenty-four hours a day.

After a long time—well, it seemed like a long time—somebody came up with this idea for a brilliant gadget, a plastic piece of Perspex with the letters of the alphabet and a hole in the middle. You hold it up between you and the other person, they look at a letter, and you can say which letter they're

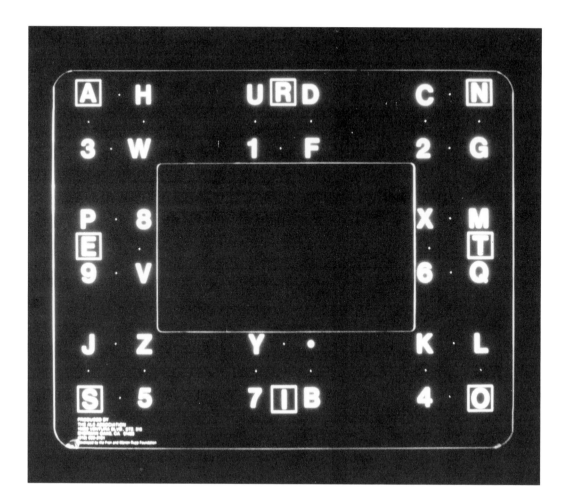

The Perspex device. For a while it was the only way Stephen Hawking could communicate.

looking at. Most of the time. Sometimes you couldn't be quite sure, so you'd get a patient to spell out what they wanted. They pick out the "A." And you say "A?" Did you get it right? It's like a guessing game.

It took a long time to get Stephen to accept the idea of using a computer to communicate. It wasn't just "Oh, I can't be bothered." It was "I don't want to do that." Stephen wasn't willing to accept that he wasn't going to speak again. He thought he was giving in by trying to find a method of communicating other than speech.

I remember I went in one evening and it was the first

154

time that he asked to be gotten out of bed to use the computer. The first thing he typed on it after saying "Hello"—Stephen was always very polite about things like that—was "Will you help me finish my book?"

STEPHEN HAWKING

*B*efore the operation, my speech was getting more slurred, so that only a few people who knew me well could understand me. I wrote scientific papers by dictating to a secretary and I gave seminars through an interpreter. The tracheostomy operation removed my ability to speak altogether.

BRIAN WHITT

*H*e had an overview of how things should work. When he couldn't say, "Look at this paper, that's where it's proven," Stephen was trying to understand the way in which the world works and saying, "That's the way it must happen because of what I understand"—not because he could prove it from what he knew at that point.

And, of course, sometimes he was wrong. And sometimes, when he told you something like that, you would go away and you'd do a calculation and come back and say, "Look, you're wrong," and he wouldn't believe you. Then you would talk about it, and a couple of weeks later you'd realize that he was right, that his hunch was better than your calculation. I think that's a very important aspect of his mind: the ability to think ahead rather than go step by step; to jump the simple calculations and just come up with a conclusion.

Both of us had a strong desire to communicate science; it's important that people who are not scientists have an understanding of what science is about. And we were both very keen to try and make this a book which people would read, although we never expected what did happen to happen.

Stephen was totally unwilling to compromise on accuracy. Of course you have to gloss over things, and of course you can't explain every detail, but as far as is possible, if you use an analogy, you don't want it to break down at the first hurdle; you would like an analogy to be sufficiently accurate that if a person starts thinking in those terms, they don't then extrapolate the analogy and find they are making nonsense of it. We spent a lot of time discussing analogies and wondering whether they really were valid. In the first draft, say, Stephen would have explained something in scientific terms and then he'd say, "Well, nobody will understand that. Let's take that one thing and come up with an analogy." Step by step, we were deliberately doing that. Now, some of the analogies in there, of course, are historic ones from Einstein or whatever, and have been used as explanations before; but some were ones which we came up with or refined.

For instance, there's an analogy to explain the complicated concept that particles which look different at lower energies might actually be the same particle at higher energies. You say, "Here we have things which are different, and we can see they're different, and we can explain how they're different—they've got different masses, different properties, and so on." And then you say, "But according to our theories, at high energies these things are the same thing; they just look different at low energies. So what does that mean?"

The analogy we came up with there was the idea of a ball

spinning on the roulette wheel. While the roulette wheel is spinning, the ball can whiz around and land on any of the numbers. So any ball which ends up in any slot on the roulette wheel is really the same. As the whole system slows down and loses energy, eventually it slots into one number—say, number 22. There's nothing special about number 22; that same ball could have ended up in any of the other slots when the energy decreased. And if you start spinning it again to increase the energy, then it'll flip around and it could be any of those numbers at all.

That was one of the analogies we came up with after sitting there and thinking, how do you explain this? What we were trying to achieve was an analogy that could be visualized.

Ian Moss was a graduate student under Stephen Hawking at Cambridge in the early 1980s. He is currently a lecturer in physics at the University of Newcastle.

IAN MOSS

Stephen would leave home in the morning, travel on his own, and arrive at work about eleven in the morning. So, of course, most of us would turn up around eleven as well. Then we'd start our coffee break. Coffee break would go on to, say, noon, and then lunch would begin about one and last until about three.

People would ask Stephen what he thought about what they were working on or anything interesting, and discussions would go on and on, if necessary all afternoon. It's a wonder that anyone got any work done because we were always talking and sitting in the coffee room all the time. I can't remember when we sat in our offices doing any work. It was a good atmosphere.

BRIAN WHITT

Stephen was a very good teacher largely because of the amount of time and interest he gave to his students. A lot of university professors are remote and they don't see their students a great deal. Stephen saw you throughout the day.

He was barely intelligible then, so there were only half a dozen people who understood him well. Then, when he lost his voice, things changed, of course; but in some ways they changed for the better, in that by losing his voice he was able to communicate with everybody.

Christopher Isham

*I*n the days when he could still speak, it was possible to understand him. I found that although I lost the art, as it were, of doing this when I was away for a long time, when I was back in his presence I could pick up what he was saying after a couple of minutes. There were occasions, certainly, when he was talking to graduate students who perhaps weren't so familiar with him, that he would get very upset because he just couldn't make them understand the word he was trying to speak. Now, of course, with the speech synthesizer, it's different.

Stephen Hawking

*F*or a time after the tracheostomy operation the only way I could communicate was to spell out words letter by letter, by raising my eyebrows when someone pointed to the right letter on a spelling card. It is pretty difficult to carry on a conversation like that, let alone write a scientific paper.

However, a computer expert in California heard of my plight and sent me a computer program which allowed me to select words from a series of menus on the screen, by pressing a switch in my hand. The program could also be controlled by a switch operated by head or eye movement. When I have built up enough of what I want to say, I can send it to a speech synthesizer, which, with a computer, is fitted to my wheelchair.

This system allows me to communicate much better than I could before. I can manage up to fifteen words a minute. I can either speak what I have written, or save it on disc.

One's voice is important. If you have a slurred voice,

159

The device Stephen Hawking uses to operate his computer program. It can be controlled by his hand, or by head or eye movements, and is known as the "clicker" because of the sound the switch makes.

The computer screen attached to the wheelchair

people are likely to treat you as mentally deficient. My synthesizer is by far the best I have heard because it varies the intonation and doesn't speak like a Dalek. The only problem is that it gives me an American accent.

Frank Hawking had become very ill while Stephen was still recovering from the tracheostomy, and died in 1986.

MARY HAWKING

My father had taken a lot of time looking into everything that had been tried or was known about motor neurone disease. One of the people who worked with him said that he virtually took a year off to work just on this, contacting everyone who might possibly know anything or be trying anything. Even today I'm still getting all the research papers on slow viruses that were being sent to him.

ISOBEL HAWKING

He was very upset by his father's death—it was rather a dreadful thing. Stephen was in hospital a very long time after the tracheostomy, and he didn't have any special sort of vehicle to take him to and fro. It was very difficult for him to come, just as it was very difficult for me to visit Stephen at that time because my husband was in need. Nobody had realized how close he was to death except Mary, who's a doctor and who was close at hand. So she phoned Stephen and told him. He hadn't realized. I think he came the next day. He was very fond of his father, but they had grown apart rather and hadn't seen a great deal of each other in the late years.

I don't think Stephen ever took any interest in his father's work, and his father never tried to interest Stephen in it, because their interests were so different. But they had a lot of common ground. He was a great inspiration to Stephen because of the compendiousness of his mind.

I think that they discussed Stephen's book in the early stages, because Stephen was writing it before his father died. I think in fact he read the first draft, and he was very interested in it.

—————————

Mechanics (either classical or quantum) does not distinguish a preferred direction of time. If you took a motion picture of the planets orbiting the sun and played it backward, the reversed motion would obey Newtonian gravitational theory just as well as the original one; it could, in principle, be the actual motion of the planets of some distant solar system.

But the universe does have a preferred direction of time. Stars burn their nuclear fuel, animals get older, and people remember the past rather than the future. Furthermore, all stars, animals, and people do so in the same direction; it is not that some of us remember the past while others remember the future. There is a universal arrow of time.

How the arrow of time arises has been understood for more than a century. Take a new pack of cards and rearrange the cards by shuffling them in any way. Do this three or four times and record the order of the cards before each rearrangement. Someone presented with the records would have no trouble arranging them in order of time. The first record is the one in which the pack is new, the cards arranged in order. The next record is one in which they are somewhat disordered. The next is one in which they are more disordered, and so on. (After four or five rearrangements, the deck gets thoroughly disordered and this procedure breaks down, but if you imagine starting with a deck of very many more than fifty-two cards, you could go for many more stages before this happened.)

This has nothing to do with the details of how the cards are shuffled. What is more important is that you begin with a new pack, one that is in a highly ordered condition. In accordance with the second law of thermodynamics, any mechanical process, not just the shuffling of a pack of cards, increases (or at least does not decrease) the disorder of a system. If the universe somehow began in a highly ordered condition, like a new deck of cards, this principle is sufficient to explain all observed instances of the arrow of time. (Although in some cases the chain of reasoning required to show this is a long and difficult one.)

Thus a complete explanation of the arrow of time requires explaining why the universe started out as it did. It is a problem in cosmology.

STEPHEN HAWKING

What would happen if and when the universe stopped expanding and began to contract? Would the thermodynamic arrow reverse and disorder begin to decrease with time? Would we see broken cups gather themselves together off the floor and jump back onto the table? Would we be able to remember tomorrow's prices and make a fortune off the stock market? I felt the universe had to return to a smooth and ordered state when it recollapsed. If this were so, people in the contracting phase would live their lives backwards. They would die before they were born and get younger as the universe got small again. Eventually, they would disappear back into the womb.

"What would happen when and if the universe stopped expanding and began to contract? Would the thermodynamic arrow reverse and disorder begin to decrease with time? Would we see broken cups gather themselves together off the floor and jump back onto the table?"

Raymond Laflamme is a French Canadian who studied under Stephen Hawking from 1984 to 1988. His work on the quantum cosmology/wave function of the universe proved that Hawking was wrong about his theory that the arrow of time would reverse when the universe contracts. He is now a Fellow of Peterhouse College, Cambridge.

RAYMOND LAFLAMME

*H*e gave me my first problem to do. And usually when he gives out a problem, he has a good idea of what the answer should be. I went to look at it and it took me a few months to understand what it was about. When I came back I said, "Stephen, I get this answer."

He said to me, "No, that is not what I expected."

But I said, "Stephen, that's what I get."

So I went to the blackboard and explained it. He said, "Did you think about this particular case?"

I said, "Oh, no, I didn't."

So I went back and calculated what he'd talked to me about, and came back a few weeks after. And I said, "Stephen, I don't get this thing. I still get the same answer I had originally."

He said, "No, no, this doesn't work. Did you think about that?" I said, "Oh, no, I'd forgotten about that particular case."

So I went back to the drawing board and started calculating again. And again I got the same answer. So I went back to see Stephen. This dragged on for about two or three months. Finally he said, "Maybe one of your approximations is not valid." So a colleague and I decided to do the thing with computers. This takes a lot of time, to write all the programs and work out the thing and be sure the program is correct.

We still got the same answer I had gotten before.

At that moment Don Page came in and he said,

"Raymond, I'm really interested by that because I get roughly the same thing. But from a different way."

So we decided we had to convince Stephen we were correct in that particular field—which was the arrow of time. I remember Don telling me, "We're better off to go slowly and convince Stephen of our assumptions before telling him the result, because if we tell him the result and it's not the result he wants, he will conclude something is wrong with our assumptions." Instead, we decided to lay down our assumptions correctly so that Stephen would agree before we told him the result. So together we worked on Stephen for about a month, and finally we convinced him we were right.

Stephen Hawking at his desk in his office

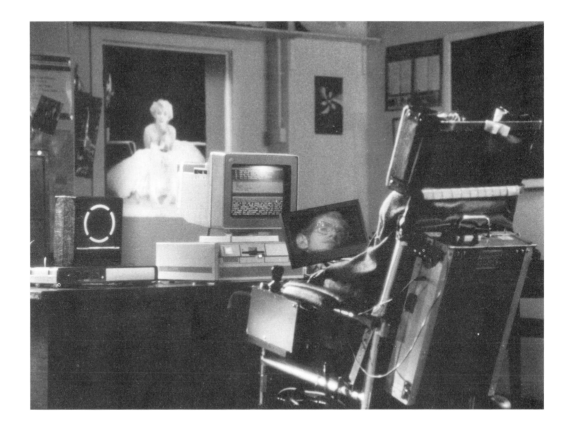

STEPHEN HAWKING

I had made a mistake. It turned out I was using too simple a model of the universe. Time will not reverse direction when the universe begins to contract. People will continue to get older, so it is no good waiting until the universe recollapses to return to our youth.

BERNARD CARR

In a sense, Stephen is forever facing up to death. It's always a possibility. I think this is an important factor in Stephen's development, in that he's determined to work fast, in that he realizes that time could be short and he must work very intensely.

But I think one thing that has kept Stephen going is his tremendous determination to understand the universe. Death is with all of us. But more so with Stephen. It does provide a strong motivating factor.

The question people always ask is, if he hadn't been disabled, would he have made such a great or an even greater contribution to science? It's not clear to me that his being disabled has actually worked in the direction of hindering his scientific output; it's possible that he might not have produced so many good results if he had not been disabled. So I would say his disability works both ways.

STEPHEN HAWKING

The universe has two possible destinies. It may continue to expand forever, or it may recollapse and come to an end at the big crunch. I predict the universe in time will come to an end at the big crunch. I do, however, have certain advantages over other prophets of doom. Whatever happens ten billion years from now, I don't expect to be around to be proved wrong.

DON PAGE

Recently, in Moscow, we were in a restaurant where they had a little dance hall and Stephen was trying to get some of us to go dancing. Well, none of us quite had enough courage or whatever to do that. But then, on our way out, we all had to pass the dance floor, and Stephen whirled around in his wheelchair there on the floor, which made quite a sight.

IAN MOSS

*T*he wheelchair is a logistical problem traveling because it has to go inside the plane—if it went into the baggage hold, then Stephen would have to hang around for an hour or two waiting for the luggage to come out, which was a big problem because there was nowhere to put him. On one occasion he was so insistent that the cabin crew agreed to put the wheelchair in a first-class seat on the plane, while we went second class. And of course it's a big problem getting him through the security checks in the airports, because anyone in a wheelchair has to go through the metal detector—and the chair won't fit.

169

Only on one occasion did the security staff insist on searching him, which I thought was a bit of an indignity. But I can imagine that Stephen presented quite a security risk. In fact, I once smuggled a box of cigars by hiding them in his wheelchair.

DON PAGE

Hawking was made a Fellow of the Royal Society—he was, at the time, the youngest. A year or two later they inducted Prince Charles, too, and they invited some of the younger Fellows to come to London. I didn't go, but Jane reported back that Prince Charles was quite enamored with Stephen's wheelchair. Stephen loves to show off and whirl it around, and apparently while whirling around, he ran over Prince Charles's toes. I hope that the Prince still has his toes all intact. That's one thing I think I have in common with Prince Charles. Both of us have been run over by Stephen Hawking's wheelchair.

STEPHEN HAWKING

On the evening of Tuesday, March 5th, 1991, at about ten forty-five, I was returning to my flat in Pinehurst. It was dark and raining. The wheelchair had front and rear bicycle lights. I came up the Grange Road and saw headlights approaching, but judged that they were far enough away that I could cross safely. However, the vehicle must have been traveling very fast, for when I just passed the middle of the road, the nurse screamed, "Look out!" I heard tires skidding and my wheelchair was struck a tremendous blow in the back.

I ended up in the road with my legs over the remains of the wheelchair. The accident destroyed my wheelchair and damaged the computer system with which I communicate. I broke my left upper arm and had cuts on my head that required thirteen stitches.

·······———·······

Forty-eight hours later, he was back at work in his office.

·······———·······

ISOBEL HAWKING

*H*e really doesn't believe he's different in any way from anybody else. If he can do it, well, so can everybody else. And that's why his book was not a specialist's book. I mean, it's addressed to anyone, and he believes that anyone can understand it. He believes that I can. I think that's a bit optimistic. But he does really believe that.

CHRISTOPHER ISHAM

*N*ormally, when doing work in theoretical physics, you have a piece of paper in front of you and you sort of scribble away, and you think a bit and you scribble away a bit more. I have a ten-to-one wastage rate of stuff that goes into my bin compared with what goes into my notes.

Now Stephen has to do all this in his head, of course, and it's very noticeable in his papers how he really does stick to the point; he doesn't get tempted to go off into side alleys, as some of us do. He uses his time very succinctly and quite simply; he always goes for the sort of simple methods where possible, which of course is the elegant thing to do in physics.

I suppose in that sense his illness has had a direct effect on the way he works. And, of course, the other obvious effect is the difficulty of communication with fellow scientists; this must be intensely frustrating for him, and for us. You can't just go up to him and say, "Look, I don't understand this. There's something in your paper, you see, this, that, and the other." Or, if you do do that, he has to set aside a very large fraction of his day to give you a proper reply. This obviously inhibits scientific communication.

We do have in theoretical physics a very efficient way of disseminating the written word: we send these preprints out to each other. And by reading what Stephen's written, of course, one gets a very good idea of what he's doing. I would say in many ways he's managed to function almost as if he didn't have the disability—you really cannot spot in his work at any stage the signs of somebody who's struggling against some terrible problem, which he obviously has, except perhaps in this clarity of style which he's developed.

I suppose that if you have any form of disability, you tend to look in upon yourself somewhat. And if you are introspective, it's certainly beneficial to theoretical physics. Theoretical physics is really a rather lonely sort of business. Although people do work together in small groups—two, sometimes even three—at the end of the day you have to sit down by yourself with your pad of paper and do your own thing.

I suppose that Stephen's disability would have encouraged him to work on those types of field where the application of intense thought was particularly appropriate, rather than just masses and masses of calculations. For example, there are branches of elementary particle physics where a piece of research can involve a very complicated calculation, either on paper or using a computer—it's a massive amount of work. Clearly, Stephen hasn't gone in for that sort of thing, for the same reason. Instead, particularly in recent years, he has tended to go for those subjects which require a lot of intense thought. There is this metaphysical aspect, and of course he does discuss this; he thinks very hard about it, about the foundations of quantum mechanics and how they relate to what he's doing. So in that sense also, perhaps one can see how his illness has maybe focused him inwards a little.

JANET HUMPHREY

I think it must be significant that the question of time comes into this so much. It must be Stephen's experience that he's somehow using in the elaboration of his ideas. The memories I have are very much kind of visual pictures of what Stephen was—of seeing Stephen in certain situations. He was always moving, always. It was the same about his face and gestures, which he used a great deal. But this is only memory. I found some photographs recently which reminded me of the general look of everybody—and I must say Stephen looked very much like he does now.

ISOBEL HAWKING

He does believe very intensely in the almost infinite possibility of the human mind. You have to find out what you can't know before you know you can't, don't you? So I don't think that thought should be restricted at all. Why shouldn't you go on thinking about the unthinkable?

Somebody has got to start sometime. How many things were unthinkable a century ago? Yet people have thought them—and often they also seem quite impractical. Not all the things Stephen says probably are to be taken as gospel truth. He's a searcher, he is looking for things. And if sometimes he may talk nonsense, well, don't we all? The point is, people must think, they must go on thinking, they must try to extend the boundaries of knowledge; yet they don't sometimes even know where to start. You don't know where the boundaries are, do you? You don't know what your taking-off point is.

JOHN WHEELER

The first time man brought down a star to earth was five-thirty in the morning of a July day, over the New Mexico desert. And that first bit of matter, in 1945, performed according to expectations and gave us marvelous evidence that we know how to predict, evidence that our theories are correct.

Here was something out of all experience of man on earth; and yet one could also say that the expansion of the universe, as predicted by Einstein, is so preposterous that in the beginning he couldn't believe it himself. It's the greatest evidence that man has ever been granted that our theories work and predict with a power beyond our original realization. We've had wrong theories, and the wonderful thing about the scientific community is that a wrong theory gets caught, and by finding out where it's wrong, we learn something new.

So, to me, the world around us is filled with what the old Romans used to call *flammanti moenia mundi,* the flaming ramparts of the world—the frontiers of human knowledge, all the way from how genes work, how life multiplies, how the universe expands, how black holes sop up information—all these things are frontiers not just of science, but of mankind itself. We are, in my view, just children at present. There's so much unexplored, so much marvelous, yet to be opened up. And I think, more and more, men, women, and children are coming to realize that we're embarked on mankind's most wonderful adventure today, in the here and the now, exploring these frontiers.

STEPHEN HAWKING

If we do discover a complete theory, it should in time be understandable in broad principle by everyone, not just a few scientists. Then we shall all, philosophers, scientists, and just ordinary people, be able to take part in the discussion of the question of why it is that we and the universe exist. If we find the answer to that, it would be the ultimate triumph of human reason—for then we would know the mind of God.

CONTRIBUTORS

Robert Berman   "He was obviously the brightest student I've ever had."

Gordon Berry   "Steve actually fell on the stairs."

Bernard Carr   "When Stephen and I were at Caltech, we had a discussion on the nature of fame."

Brandon Carter   "It would be a very exciting way to end one's life."

Michael Church   "Suddenly I was aware that he was egging me on."

Norman Dix   "Some coxes can be adventurous and some coxes can be very steady."

Jim Hartle   "Anyone over the age of twelve knows there is no such thing as certainty, right?"

Edward Hawking   "But it was our house and we loved it anyway."

Isobel Hawking   "Luck. Luck. Well, we have been very lucky."

Mary Hawking   "I always had the impression that fathers were like migratory birds."

Stephen Hawking   "They asked me about my future plans. I replied I wanted to do research."

Antony Hewish	"Who would dream you would get intelligent-looking signals coming from the sky?"
Sir Fred Hoyle	"I prefer to tackle problems which I can see are soluble."
Janet Humphrey	"He once decided it would be nice to have Scottish dancing in the evening."
Christopher Isham	"You have this nice picture of no past—all you can really say is that the universe *is*."
Basil King	"We bet a bag of sweets on the issue."
Raymond Laflamme	"Stephen, I still get the same answer I had originally."
John McClenahan	"It was the kind of family that did those sorts of odd things."
Ian Moss	"I once smuggled a box of cigars by hiding them in his wheelchair."
Don Page	"That is a question that science can neither affirm nor refute."
Roger Penrose	"I had had some idea in crossing the road, but then it got completely blotted out of my mind."
Derek Powney	"I was sitting there one evening with him when he asked, 'Have you ever read John Donne's elegies?'"
Patrick Sandars	"By that time it was clear that Stephen knew more about the subject than I did."
Dennis Sciama	"After the first step we couldn't stop the progress. Almost every year an exciting new discovery was made."

John Taylor	"Of course, your wristwatch is destroyed; so is your wrist, sadly enough."
Kip Thorne	"He developed a very powerful set of tools that nobody else really had."
John Wheeler	"The girl is the ordinary star and the boy is the black hole."
Brian Whitt	"The analogy we came up with was a ball spinning on the roulette wheel."

AFTERWORD

A NOTE ON THE FILM AND THE BOOK

·······———◆———·······

Stephen Hawking's A Brief History of Time: *A Reader's Companion* was created to give the reader of *A Brief History of Time,* and the viewer of director Errol Morris's documentary film adaptation of that book, a clearer insight into Stephen Hawking's life and work. It is comprised in large part of materials gathered for the film, which is also called *A Brief History of Time,* including interviews, illustrations, and Professor Hawking's own narrative.

In the Acknowledgments to his book, Hawking wrote of his conviction that basic ideas about the origin and fate of the universe can be stated without mathematics in a form that people without a scientific education can understand. He accomplished this with dazzling success, as the book's worldwide popularity attests. Yet in another sense the book is wonderfully deceptive; Stephen Hawking's gift and wit as a writer disarm the reader, and its powerful concepts invite further questions about the science itself and about the author.

The film and this book have been created to explore further both of these aspects. The evolution of *A Reader's Companion* is a journey not only through Professor Hawking's world but through the world of documentary filmmaking. The medium of film allows viewers to absorb images of people and complex ideas visually and voices and music through sound, and it is the filmmaker's challenge to shape and present this content so that it is most effectively realized. The printed page offers a different challenge and makes different demands. It is our hope that in their differing ways the film and the book are responsive to the eagerness of viewers and readers to learn more about Stephen Hawking and his work.

The film project took nearly three years, and provided a rare and stimulating opportunity: to seek out those people who form Stephen Hawking's personal and professional universe and incorporate their insights, opinions, and recollections into an integrated portrait; and, of course, to hear from Hawking directly and at length.

When we first approached Professor Hawking with the request to make a film of *A Brief History of Time,* we felt an understandable anxiety about proposing a collaboration between filmmakers and a noted scientist. But after the initial set of meetings, that unease turned into a very strong working relationship. In the thick of the editing, Stephen Hawking and Errol Morris could be seen in the edit room for hours on end working toward a single vision for the film.

The production brought together not only Hawking and several members of his family, but colleagues, lifelong friends, and a number of the world's top physicists. Morris's uncanny ability to get people to reveal themselves in the process of telling a story or describing their work delivered thousands of feet of filmed interviews. This material was so rich that it quickly became evident a book should be developed from it. Because a documentary like this, with interviews comprising the heart of the narrative, runs only ninety minutes, it could utilize only a small portion of what was actually recorded on film. In addition, Professor Hawking wrote a great deal of new narration that could not be incorporated for reasons of length.

This book takes advantage of these hundreds of hours of interviews, including those of nine people who could not be included in the final film, and allows Stephen Hawking to speak more fully and directly about his life and his work. The book also contains background information about the participants and the scientific concepts, family pictures, and artwork by Ted Bafaloukos beyond that used in the film.

In its faithfulness to the underlying concept of the film as a journey through the profoundly important work of Stephen Hawking and the life that has shaped that work, this book also seems to the filmmakers to be a true documentary. It is the work of many people beyond those who created the film. Linda Grey, president and publisher of Bantam Books, and Ann Harris, senior editor at Bantam, deserve special thanks for their patience, diligence, and insight. Appreciation goes to Gene Stone, who prepared the text, and Michael Mendelsohn, who designed the book, for their attentive care. Al Zuckerman, the book's literary agent, has been a faithful steward to the project. The twenty-seven people who participated in Errol Morris's interviews are owed thanks not only for their time but for the candor and illumination of the contributions that appear in the film and throughout these pages. The help of Sidney Coleman, the science adviser to the film, was invaluable. Special thanks go to David Hohmann, the graphics coordinator of the film, who helped gather the book's illustrations. Very special thanks are due Stephen Hawking's staff: Sue Masey, Andrew Dunn, Stuart Jamieson, and Jonathan Brenchley.

Above all, Errol Morris, Colin Ewing, executive producer for Anglia Television, David Hickman of Anglia Television, the producer of the film, and I are indebted to Stephen Hawking. We are grateful for the opportunity to interpret his work on film, and for the wit, warmth, and enthusiasm with which he participated in this shared enterprise of film and book.

GORDON FREEDMAN
Executive Producer
A Brief History of Time

GLOSSARY

antiparticle Every type of element particle has an antiparticle of the same type. When a particle meets such an antiparticle, they annihilate, leaving only energy.

atom The building block from which everyday matter is made. Atoms are composed of a nucleus of protons and neutrons which is orbited by electrons.

big bang The singularity at the beginning of the universe, when all of the universe was at a single point of infinite density and temperature.

big crunch The singularity at the end of the universe, when all of the universe collapses to a single point of infinite density and temperature.

black hole A region of space-time from which nothing can escape because gravity is too strong. Even light travels too slowly to escape; hence the region does not emit any radiation and appears black. However, the uncertainty principle of quantum mechanics allows particles and radiation to leak out of a black hole.

classical mechanics A system of laws in which each object has a definite position and speed. It has now been succeeded by quantum mechanics, in which objects do not have definite positions and speeds.

cosmic ray A high-energy particle of matter, from space, that travels at a speed close to that of light.

cosmology The study of the universe in its entirety.

electron	An elementary particle that usually orbits the nucleus of an atom. It belongs to a family of low-mass matter particles called leptons ("light ones"), and has an electrical charge of −1.
elementary particles	Particles that do not have any internal structure. They fall into the categories of matter particles or force-carrying particles. Each type of particle has an associated type of antiparticle.
entropy	A measure of the disorder of a system, which according to the second law of thermodynamics must always increase.
event horizon	The boundary of a black hole. Once this boundary is crossed, it is impossible to escape from the black hole.
frequency	For a photon, the rate at which the electromagnetic field associated with the photon changes. The higher the frequency, the greater the energy of the photon.
gamma ray	A photon of particularly high energy, which could be emitted by a nuclear reaction or a low-mass "primordial" black hole formed in the early stages of the universe. A typical gamma ray will have a wavelength of about 0.0000000001 meter.
general relativity	Einstein's second theory of relativity (1916), which states that gravitation results from distortions in space-time geometry (that is, geometry that considers distances not only between points in space, but between points in space and time), and which established that gravitational fields affect measurements of time and distance.
Hawking radiation	The elementary particles and radiation emitted from the event horizons of black holes. The smaller the black hole, the greater the amount of Hawking radiation and the faster the black hole shrinks, leading to a vast explosion as the black hole finally evaporates and disappears.

imaginary time	An idea in which the time variable in equations is treated as an imaginary number, i.e., a number which is a multiple of the square root of −1.
inflation	A period of accelerating expansion that is thought to have occurred in the very early universe.
microwave	Radiation with a wavelength of about 1 centimeter.
microwave background radiation	A radiation in the microwave region of the electromagnetic spectrum which propagates uniformly through the universe in all directions. This background radiation is a remnant of the enormous heat triggered by the big bang and is therefore considered a confirmation of the theory.
neutron	A non-elementary particle with no electrical charge, usually found in the nucleus of an atom. It is composed of elementary particles called quarks.
neutron star	A star that has become so dense that its gravity is strong enough to cause most of the electrons and protons in the atoms of the star to combine and become neutrons.
no boundary proposal	The proposal that space and imaginary time together form a surface that is finite in extent but does not have any boundaries or edges. In this proposal, space-time would be like the surface of the earth, but with two more dimensions.
optical telescope	A telescope that forms images of the stars and galaxies from light visible to human beings.
photon	An elementary particle or quantum of light.
primordial black hole	A black hole formed shortly after the big bang.
proton	A non-elementary particle usually found in the nucleus of an atom. It has an electrical charge of +1, and is composed of elementary particles called quarks.

pulsar	A rotating neutron star that emits pulses of radio waves as its magnetic field interacts with the magnetic field surrounding it.
quantum gravity	A theory that unites quantum mechanics and general relativity.
quantum mechanics	A system of theories in which particles do not have exactly defined positions and speeds but behave like waves in many respects. Similarly, waves such as light behave like particles in many respects.
quasar	A starlike object thought to consist of an enormous rotating black hole with large amounts of matter falling into it. The matter gets very hot and emits a lot of energy before it falls inside the black hole. Quasars are extremely distant, but can be observed because they are so powerful.
radio telescope	A telescope that makes maps of the positions in the sky of sources of radio waves such as quasars and radio-emitting galaxies.
radio waves	Waves in the electromagnetic field that are like waves of visible light, but have much longer wavelengths, of the order of meters rather than centimeters.
second law of thermodynamics	A law that says that the amount of disorder, or entropy, in the universe increases with time. It is different from other laws in that it is not always true, but only nearly always, and it depends on the universe's starting out in an ordered state.
singularity	A point of infinite curvature of space-time at which space-time comes to an end. The classical general theory of relativity predicts that singularities will occur, but cannot describe what happens at a singularity because the theory breaks down here.
space-time	The four-dimensional description of the universe from

relativity, uniting the three space dimensions and the single time dimension. Curved space-time is used in the general theory of relativity to describe gravity.

special relativity
Einstein's first theory of relativity (1905), which says that light always travels at a constant speed, and that the speed of light is an absolute constant, no matter where it travels. This theory unified space and time into a flat, four-dimensional space-time, but did not describe the effects of gravitation.

steady state theory
A theory of cosmology, now largely discredited, in which new matter is formed in the expanding space between existing galaxies.

thermodynamics
The branch of physics concerned with heat and the other forms of energy.

uncertainty principle
The principle which states that one can never be exactly sure of both the position and the velocity of a particle; the more accurately one knows the one, the less accurately one can know the other.

virtual particle
The uncertainty principle of quantum mechanics allows the amount of energy in the universe to fluctuate about a fixed total for brief periods of time. The larger the fluctuation, the shorter its allowed duration. Particles created from such energy fluctuations are called virtual particles, and annihilate when energy must be paid back.

wave function
A distribution that describes the probability of finding a particle at different points.

wave function of the universe
A distribution that describes the probability of finding different shapes for the universe at a certain time.

white dwarf
A star that has reached a stable state, in that it is not massive enough for gravity to become sufficiently strong to force it to collapse in on itself.

INDEX

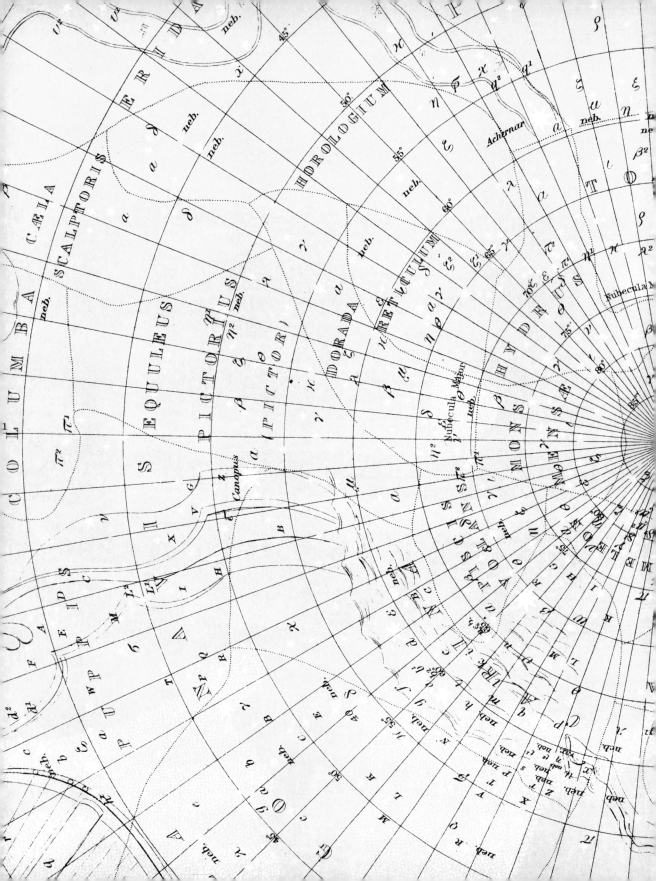